Challenges of the Agrarian Transition in Southeast Asia (ChATSEA)

Powers of Exclusion
Land Dilemmas in Southeast Asia

Challenges of the
Agrarian Transition in Southeast Asia
(ChATSEA)

The shift from rural societies dependent upon agricultural livelihoods to predominantly urbanized, industrialized and market-based societies is one of the most significant processes of social change in the modern world. The Challenges of the Agrarian Transition in Southeast Asia (ChATSEA) project examines how the transformation is affecting the societies and economies of Southeast Asia. Headed by Professor Rodolphe De Koninck, holder of the Canada Chair in Asian Research at the University of Montreal, and sponsored by the Social Sciences and Humanities Research Council of Canada, the ChATSEA project includes publications by senior academics as well as junior scholars.

Powers of Exclusion

Land Dilemmas in Southeast Asia

Derek Hall
Philip Hirsch
Tania Murray Li

NUS PRESS
SINGAPORE

NUS Press
National University of Singapore
AS3-01-02, 3 Arts Link
Singapore 117569

Fax: (65) 6774-0652
E-mail: nusbooks@nus.edu.sg
Website: http://www.nus.edu.sg/nuspress

ISBN 978-9971-69-541-5 (Paper)

National Library Board, Singapore Cataloguing-in-Publication Data

Hall, Derek, 1971-
    Powers of exclusion : land dilemmas in Southeast Asia / Derek Hall, Philip
Hirsch and Tania Murray Li. – Singapore : NUS Press, c2011.
    p. cm.
    Includes bibliographical references and index.
    ISBN : 978-9971-69-541-5 (pbk.)

    1. Land tenure – Southeast Asia. 2. Land use, Rural – Southeast Asia. 3.
Land reform – Southeast Asia. 4. Agriculture – Economic aspects – Southeast
Asia. 5. Agriculture – Social aspects – Southeast Asia. I. Hirsch, Philip, 1957-
II. Li, Tania, 1959- III. Title.

HD880.8
333.3159 -- dc22   OCN696693199

Typeset by: Forum, Kuala Lumpur, Malaysia
Printed by: Mainland Press Pte Ltd

Contents

List of Figures

List of Tables

Acknowledgements

The impetus to write this book came out of our participation in the research programme "Challenges of the Agrarian Transition in Southeast Asia" (ChATSEA), which was funded by Canada's Social Sciences and Humanities Research Council between 2005 and 2010. We would like to thank the other participants in the programme, faculty colleagues and graduate students alike, for their camaraderie, for their comments and suggestions, and for expanding our understanding of agrarian Southeast Asia. We are especially thankful to Rodolphe De Koninck (the project director), Bruno Thibert, Monia Poirier and Bruno Gendron for their work at "ChATSEA Central" at the Université de Montréal. We would not have been able to write the book without meeting together on several occasions, and we are grateful to Nancy Peluso for letting us use her office in Berkeley for our meetings, to University College at the University of Toronto for ten days of accommodation and working space at the incredible Bissell House, and to ChATSEA for funding our travel. We received terrific feedback on drafts of parts of the book at a ChATSEA workshop in Montréal; at the October 2007 meetings of the Canadian Council of Southeast Asian Studies at Université Laval; from Nancy Peluso's Land Lab group in September 2009; and from Rob Cramb, Josephine Gillespie and an anonymous reviewer at NUS Press. Elissa Janca provided excellent research assistance at the beginning of the project, Sunandini Arora Lal of NUS Press did a painstaking job of copyediting and proofreading, and Bruno Thibert generated the maps. Derek Hall would also like to thank Tanya Richardson for her wonderful advice, encouragement and support during the writing process.

The text is the product of our joint and equal authorship, developed through weeks of intensive discussion and months of writing and rewriting, and sustained by the rewards of a collaboration that challenged each of us to push ourselves beyond our usual disciplinary, geographical and theoretical comfort levels. That we learned a huge amount from each other, and produced something none of us could even have conceptualized, still less written alone, is something we're very pleased to acknowledge.

1

Introduction

Setting the Scene

The period since the 1980s has seen dramatic yet seemingly incongruous changes in rural land use and social relations around land in Southeast Asia. Some of the main processes driving these changes seem to represent threats to agriculture. Economic growth, industrialization and urbanization have prompted the conversion of a great deal of agricultural land to commercial, industrial, residential, tourist and infrastructural use. At the same time, heightened concern for environmental conservation has led state actors to zone huge amounts of land as off limits to agriculture and, increasingly, to try to enforce that zoning. The disappearance of so much once-farmed land must be understood as a core component of the "deagrarianization" of Southeast Asia, that is, the process by which agriculture becomes progressively less central to national economies and to the livelihoods of people even in rural areas.

Processes working against agriculture, however, coexist with others that speak to the enduring importance of farming. Most strikingly, the area of land dedicated to crops and livestock continues to expand rapidly, mainly at the expense of forest, and much existing farmland is being used more intensively. National governments across the region have embarked on ambitious programmes to provide formal property rights in land to smallholders. They have been inspired to do so, in large part, by national and transnational discourses that put agricultural land at the heart of economic development. Arguments about deagrarianization also sit awkwardly alongside popular mobilizations under a range of banners (those of the poor, the citizen, the political party, the peasant, the ethnic or indigenous group) for access to land as a productive resource, as a backstop for precarious livelihoods, and as a symbol of identity and belonging. These tensions between agriculture's retreat and its enduring importance, moreover, are not simply macro-level features of

Southeast Asian societies. They are also visible in the behaviour and stated desires of key actors. Government policies operate to both encourage and inhibit agriculture, while rural people fight for access to farmland and to ethnic homelands even while they absent themselves from the countryside to pursue "post-agrarian" futures.

A brief look at one Southeast Asian country, Cambodia, gives a sense of the extent and rapidity of change in land relations. In 2005 agriculture made up 33 per cent of GDP, and 68 per cent of the population drew their livelihoods from agriculture. Thus far, farming has been mainly smallholder-based, but in the decade 1998–2008 around 100 plantations ranging in size from 500 to 330,000 hectares were granted licences to produce crops such as pulpwood, rubber and cassava. Non-agricultural land uses also expanded: 23 permits were issued to develop "special economic zones", 88 mining corporations had concessions, 67 hydropower schemes were under consideration, and 18 per cent of the land mass was zoned as "protected" and off limits to cultivation of any kind.[1] A host of community-based programmes sponsored by NGOs supported the declaring of land as off limits or restricted for agricultural, forest, fishing and other forms of use. At the same time, the World Bank was sponsoring a land-titling programme that ran into criticism on two fronts. Advocacy groups argued that the programme was failing to provide adequate protection to groups vulnerable to eviction, while the government argued that titling hampered development schemes that required evictions in order to proceed. These schemes included several private ventures of government officials and members of the Armed Forces.[2] Rather than respond to complaints from, and conditions imposed by, the World Bank, the government cancelled the programme in 2009. Meanwhile, smallholders throughout Cambodia were taking their own initiatives to consolidate their claims to land for livelihood security and also, in some cases, as the means to a decent income from new cash crops. Their goal was not only to defend their land against the state and corporations, but also to respond to the actions of kin and neighbours who were consolidating their own claims, and some of whom were accumulating land.[3] Finally, there were moves by groups in northeastern Cambodia who had been displaced through years of war, Khmer Rouge rule and sundry development schemes to reclaim ancestral ethno-territories, or to migrate into the territories of other ethnic groups in search of land and livelihoods.[4]

Cambodia is perhaps extreme because of the compression of many sources of change into a short period of time, and a relatively small space, but elements of these processes are found across the Southeast Asian region. Taken together, the processes we explore in this book are significantly and, for the most part, irreversibly shifting the relations between people and land in Southeast Asia,

Table 1.1 Southeast Asia: Total population, agricultural population and value added by agriculture to GDP

	Total population[1] (millions)		Agricultural population[1] (millions)		Agricultural population as per cent of total population		Agriculture, value added, as per cent of GDP[2]	
	1980	2005	1980	2005	1980	2005	1980	2005
Cambodia	6.8	13.9	5.1	9.4	76	68	47 (1993)	33
Indonesia	146.6	219.2	78.6	89.1	54	41	24	13
Laos	3.2	5.9	2.6	4.5	80	76	60 (1990)	45
Malaysia	13.8	25.6	5.4	3.7	39	15	23	8
Philippines	48.1	85.5	24.7	31.1	51	36	25	14
Thailand	47.3	65.9	30.4	29.9	64	45	27	9
Vietnam	53.3	84.1	39.1	54.9	73	65	40 (1985)	21
Total	319.1	500.1	185.9	222.6	58	45	–	–

Notes: [1] FAOSTAT. The FAO defines agricultural population as "all persons depending for their livelihood on agriculture, hunting, fishing and forestry. It comprises all persons economically active in agriculture as well as their non-working dependents."
[2] World Development Indicators. 1980 data is not available for Cambodia, Laos and Vietnam.

and shaping the course of future developments. They are doing so in a context in which large absolute numbers of people continue to live in rural areas and draw their livelihoods from agriculture—37 million more people in 2005 than in 1980—even as agriculture's relative importance to GDP declines (see Table 1.1).

Our aim is to explore how and why all this is coming about. What are the powers at work in these transformations? Who are the actors promoting or opposing these changes in land relations? What dilemmas and debates do the changes invoke and provoke? Who wins, and who loses, at different sites and scales across this variegated terrain?

Exclusion

To guide our inquiry into this complex field, we focus on the changing ways in which people are excluded from access to land. The term "exclusion" is widely used in studies of land access around the world, and its use tends to have two characteristics. Empirically, references to exclusion as a *condition* tend to denote situations in which large numbers of people lack access to land or in which land is held as private property, while references to exclusion as a *process* highlight large-scale and often violent actions in which poor people are evicted from their land by or on behalf of powerful actors. Normatively, exclusion is seen as negative and is counterposed to a positively weighted "inclusion". These framings convey the sense that exclusion is something imposed on the weak by the strong, something that should be opposed.

Our approach to exclusion is different. While the chapters that follow do discuss many instances of exclusion that involve glaring inequality and dispossession, our analysis is based on the observation that all land use and access requires exclusion of some kind. Even the poorest people, farming collectively and sustainably, cannot make use of land without some assurance that other people will not seize their farms or steal their crops. Starting from the assumptions that exclusion is inevitable and that all political perspectives on land relations—from the most conservative to the most radical—take some types of exclusion to be positive, we focus on the changing means by which different social groups have gained or lost access to land in the 1990s and 2000s, and on the implications of exclusion for rural social relations.

Exclusion is not a random process, nor does it occur on a level playing field. It is structured by power relations. Across rural Southeast Asia and elsewhere, exclusion from land can be understood in terms of the interaction between *regulation, force, the market* and *legitimation.* Regulation, often but not exclusively associated with the state and legal instruments, sets the rules regarding access to land and conditions of use. Force excludes by violence or the threat of violence, and is brought to bear by both state and non-state actors. The market is a power of exclusion as it limits access through price and through the creation of incentives to lay more individualized claims to land. Legitimation establishes the moral basis for exclusive claims, and indeed for entrenching regulation, the market and force as politically and socially acceptable bases for exclusion. These powers of exclusion constitute both the title and the analytical core of the book.

Our investigation of the intersection of these powers focuses on *processes* and *actors*. The bulk of the book is divided into six chapters, each of which examines one of the key processes driving changes in rural land relations. These processes are, in order: (1) the regularization of access to land, through state-backed programmes of land titling, formalization and settlement; (2) the spatial expansion and intensification of efforts to conserve forests by restricting agriculture; (3) the arrival of new "boom crops" that see the massive, rapid and hard-to-reverse conversion of land to monocropped production; (4) the conversion of land to post-agrarian uses; (5) processes of agrarian class formation on an "intimate", village scale; and (6) the mobilization of collectivities to defend or assert their access to land, at the expense of other land users and other land uses. Taken together, these processes highlight the changing—rather than declining—significance of rural land. This is true with respect to land's status both as a productive asset that is central to livelihoods, and as territory loaded with affective significance.

One indication of the profound changes that have taken place in land relations in Southeast Asia is that the processes explored in this book are far more diverse than those that would typically have been addressed in a book about rural land relations in the "green revolution" decades of the 1960s, 1970s and 1980s. This earlier generation of work focused mainly on patterns of differentiation at the village level (our process #5) and the national policy context that promoted or inhibited differentiation at this scale.[5] The range of *actors* involved has correspondingly increased. In addition to landlords, tenants, moneylenders and tax collectors (the "local powers" examined in classic village studies), the actors we describe include smallholders and large estates. They include various central and local government agencies and functionaries, whose diverse and often contradictory agendas, rules and enforcement procedures complicate the notion of a singular "national" policy context. They expand beyond the nation to include international NGOs and donors who lend a transnational governmentality to the powers of exclusion under discussion. They also include nationally based and more locally based social movements, and a plethora of agents under the broad umbrella of "civil society". Transnational and national corporations also materialize the powers of exclusion, and indeed the lure of foreign investment often stimulates new kinds of exclusion. In practice, these actors often blur into one another. Attempts to categorize actors are complicated, for instance, when state agents employ official authority for their own private purposes or when donors, firms and NGOs work together on projects with exclusionary potential.

Exclusion is not new. We wish to avoid a crude "before and after" approach, and eschew any romanticized or simplistic notions of a lost past of local communities whose wholly inclusive land relations have been destroyed by capitalism, modernity and the state. Similarly, the dynamics we cover are often in tension with one another and cannot be boiled down to a single story. Nevertheless, the material covered in this book does point towards certain qualitative shifts, identifiable directions, intensifications and distinctive breaks with pre-existing circumstances. We identify five of these. One involves moves from more flexible and overlapping to more rigid and clearly bounded definitions of access to land. Another sees rules and norms of access based in kinship and other locally constituted social relations shifting to more state-centred and formalized means of establishing the right to exclude. A third is the partial or, in some areas, full closure of the land frontier that has long served to accommodate the needs of new generations, and served also as an escape route for people facing oppression or bankruptcy. Fourth, the transnational component of exclusionary dynamics in rural Southeast Asia, exemplified by the actions of the World Bank and international NGOs, has become much more influential in recent decades. Finally, the relative importance of different legitimations has shifted, a process most visible in the rise of conservation, formalized private property and ethnicity/indigeneity as justifications for exclusion.

New forms of exclusion involve changes in the ways in which the use of land is reserved for some actors and denied to others. These changes are often resisted by the people who experience them first-hand, though this is not always the case. Excluded people may acquiesce, protest or simply disappear in ways that make exclusion's impacts extremely difficult to ascertain. There is also a large body of literature critiquing exclusion, one that focuses on high-profile cases such as land titling, national park establishment, plantations and tourist resorts. The critique of exclusion, however, is often problematic. In the next section, we explain why this is so by identifying a key dilemma that arises from what we call exclusion's double edge: the fact that exclusion is the normal rather than the exceptional state of affairs, and that the very widespread aspirations for access to land implicitly include the wish for a degree of exclusionary power. Before we can unpack this dilemma, however, we need to specify more fully what we mean by exclusion.

Exclusion's Double Edge

There are two main strands of work related to exclusion in the literature on land and other natural resources. In economics, resources and other goods and services are said to be "excludable" when it is possible to prevent people from having access to them, while they are non-excludable when this is not possible. Excludability is thus partly a characteristic of the resource itself, but also derives from the social relations that surround it.[6] A classic example of a non-excludable good is the air we breathe. Because people can be excluded from accessing or benefiting from land (for instance, by fencing it, posting guards around it, or simply making it known through signs or public knowledge and associated sanctions that someone has claimed it), land is an excludable resource by this definition. Exclusion is also frequently discussed by critical political economists writing on natural resources, but in ways that tend not to be clearly defined or specified. We would suggest that exclusion as applied to land carries at least two (usually implicit) meanings in this literature. At times, it refers to a condition in which the distribution of land is highly skewed, with substantial numbers of people not having access to or owning any while a few hold large amounts. This meaning, which is quite close to the concept of inequality, underpins Jun Borras and Jennifer Franco's description of the "status quo" in many countries as "a condition that is marked by inequity and exclusion in land-based social relations".[7] Other scholars, however, use exclusion in a way that seems closer to the concept of private property than to inequality. Thomas Sikor and Tran Ngoc Thanh, for instance, describe what they call "exclusive forest devolution" in Vietnam's Dak Lak province as a policy that "granted local actors ownership-type property rights to forests, including the right to exclude others and a limited right of alienation". Sikor and Thanh refer to the process by which this policy was imposed as "enclosure", and the alternative "inclusive" policy they recommend involves the recognition of customary rights and the creation of forms of governance that include all relevant actors.[8] What unites the uses of exclusion by critical scholars is less the specific meanings they attach to the concept than their view of exclusion as something that is both *negative* and *ameliorable*. Exclusion is something that can, in principle, be overcome or eliminated and that can be replaced with "inclusive" relations.

Our approach differs in that we take the opposite of exclusion to be not inclusion but *access*. Building on Jesse Ribot and Nancy Peluso's definition of access as "the ability to benefit from things",[9] we define exclusion as referring to the ways in which people are *prevented* from benefiting from things (more specifically, land). We divide processes of exclusion into three main types: the ways in which already-existing access to land is maintained by the exclusion of other potential users; the ways in which people who have access lose it; and the ways in which people who lack access are prevented from getting it. Like Ribot and Peluso's concept of access, our understanding of exclusion is broader than the concept of property, even when understood broadly as involving "some kind of *socially acknowledged and supported* claims or rights": it refers not just to the presence or absence of rights but to the broader array of powers that prevent people from benefiting from land.[10] However, our prioritization of exclusion rather than access does not simply involve producing a negative image of the mechanisms of access discussed by Ribot and Peluso. A focus on the people who are kept out and the powers that keep them out involves, for instance, paying more attention to contention and conflict, a point brought home by the way force (one of our four powers) plays a much more prominent role in our book than in Ribot and Peluso's article.

Defining exclusion in this way helps to focus attention on exclusion's double edge, and the intricate problems that follow from it. Exclusion, as we have already noted, is a necessary feature of every type of land use and tenurial arrangement. Conserving a forest means keeping it off limits to (at least some) agriculture. Farmers would not raise crops without some assurance that they can hold on to their fields until harvest. Long-term investments in land depend upon the right to exclude others from interfering with enjoyment of the benefits of that investment, though they may be permitted to gather or glean. Hence, exclusion creates both security and insecurity. From the moment land becomes scarce, the exclusive access to land that is productive for some comes into tension with the fact that others cannot access it. This tension lies at the heart of land struggles both "on the ground", where people asserting contradictory claims confront each other, and in the domain of national and transnational policy, where optimal arrangements for land access are debated. Exclusion's double edge means that the various kinds of exclusivity that different actors want generally bring with them not just the desired positive effects but also a series of other effects that are much less welcome. While policymakers, NGOs, smallholders, the landless and others often state their preferences with regard to land with confidence and without acknowledging the "downsides" of their proposals, this does not necessarily mean they are unaware of them. Indeed, many of the long-standing tensions and contradictions in the politics of land in Southeast Asia stem from the fact that participants in these debates recognize exclusion's double edge but cannot identify a solution to the dilemma it presents.

Exclusion's double edge raises troubling and perhaps insoluble problems for specific people and groups. Policy debates, for example, tend to be biased towards the benefits of exclusion as the means to achieve particular ends. Experts argue that forest conservation is necessary for the future of the planet, and that people must make way for dams to enable electricity production. The World Bank, among others, argues that creating a market in land maximizes access by the parties who use land most efficiently to secure economic growth. Yet policy prescriptions advocating exclusion are often accompanied by concern over what exclusion means for human livelihoods. Colonial authorities in Indonesia, for example, passed laws that enabled them to appropriate massive areas to allocate to Europeans for plantation agriculture. But the same authorities simultaneously created a category of "native land" that they made communal and non-alienable. They were afraid that individualized landholdings would enable "lazy" natives to mortgage or sell up, placing their livelihoods at risk. For some officials, the concern was native welfare; for others, it was the risk that dispossessing too many natives too fast would disrupt both order and profit. Contemporary authorities often have similar concerns and devise schemes to restrict land alienation among groups thought to need special protection.[11]

When state authorities have recognized exclusion's double edge, and acted to try to limit the damaging effects of otherwise welcome forms of exclusion, they have been part of "countermovements". Land, as Karl Polanyi observed, is not an ordinary commodity but the basis of life, and hence exclusion from access to land is continuously subject to what he calls countermovements recalling land's social function.[12] Countermovements do not have a singular source or rationale, nor do they represent a coherent set of interests. They are loose assemblages in which diverse actors, ideals and imperatives come together provisionally, fragment and realign. Polanyi emphasized market-based exclusion as the primary threat to land access. Countermovements concerned with restricting the commoditization of land have a long history in Southeast Asia, and they have continued to emerge and reassemble in the neo-liberal era in which markets appear to be triumphant. At the same time, the scope of countermovements around land has expanded with the recognition that it is not only market processes but also regulatory regimes and land-grabbing that destroy the conditions for the maintenance of some people's lives and livelihoods. Rural people who occupy a patch of land to grow food act in terms of a countermovement when they disregard official forest boundaries or signs that say "Private property. Keep out". Their actions may be supported by activists or officials quoting clauses in national constitutions stating that the land belongs to "the people", who are entitled to use it for their survival. Also party to countermovements are ruling regimes that take land from private owners to give to other people through programmes of land reform, or allocate public land for smallholder use. There are transnational elements in these assemblages, when rural people struggling for land justify their claims in the name of the right to food promised in transnational human rights instruments. In pursuing all of these objectives, participants in counter-movements tend to frame exclusion as bad in the same way that policymakers tend to see it as good. Yet countermovements do not resolve the dilemma: even when they define the protection of life as their goal, the regimes of access they propose routinely require new forms of exclusion that benefit some parties at the expense of others.

Exclusion's double edge and the dilemma it presents to different actors come through again and again in this book. Conservation is one arena in which exclusionary effects are obvious. Protected areas enclose land and place it off limits for smallholder agriculture. The creation of protected areas may be accompanied by eviction. Compensation, if it is given at all, is seldom sufficient to replace the livelihoods smallholders lose, and smallholders often contest their exclusion in the name of their right to land as a source of livelihood. One way of looking at these struggles is in terms of scale. Actors excluded from direct access to protected areas are asked to pay the price for a global benefit, the conservation of biodiversity. That the exchange is so unfair is a symptom of the highly unequal powers of the parties involved. More significantly, the actors promoting conservation do not see their work as a land-grab—they see it as part of a countermovement for the protection of life. They view themselves as protecting forest from logging or conversion to agriculture not for their own private gain, but to secure the health of the planet and the livelihoods of future generations. They are distressed to find themselves branded as villains who care nothing for people.

Capitalism presents a dilemma that runs deeper still, because it has chronic, exclusionary effects at its core. To explain these effects, scholars working in the Marxian tradition highlight the crucial difference between producing for the market when a lucrative opportunity arises, and being *compelled* to produce for the market.[13] It is the element of compulsion that defines capitalist relations. Participation in markets is ancient and ubiquitous in Southeast Asia; the difference is that very few, if any, farmers in the region today can sustain themselves *outside* market relations. Limited access to land, combined with the need for cash to pay for the goods and services rural people widely agree are now necessary (such as schooling), compel farmers to participate in markets as producers of crops, or sellers of labour. If their costs of production (land, labour and capital) exceed the returns, or their income is insufficient to sustain the family, they fall into debt and risk losing their land. Meanwhile, other farmers, often described as more efficient, accumulate both land and capital. Efficiency and the "natural" operation of market forces in a context of competition should not be overstated: in all actually existing capitalisms, states intervene to set conditions in which selected groups are enabled to prosper and others are dispossessed, a point we develop throughout this book. Nevertheless, agrarian capitalism—like capitalism in general—systematically produces wealth and poverty, accumulation and dispossession.

Remarkably, the depth of capitalism's exclusionary effects often goes unacknowledged by development experts of a neo-liberal persuasion who advocate intensified market dependence as a means to reduce poverty, neglecting to mention the poverty that is co-produced with wealth. Alternatively, they may recognize the poverty but highlight the benefits awaiting the expelled population when they move to town, or envisage the provision of "safety nets", or shift scales to observe the public benefits of economic growth in terms of the greatest good for the greatest number.[14] Or finally, they may seek to mitigate the problem by dividing the population, separating potentially proficient market subjects from social groups whose special character—as indigenous people, beneficiaries of land reform, forest dwellers or members of other specially designated "communities"—indicates that they should be protected from the risk of market exposure. In the Southeast Asian region, the indigenous rights movement and the movement promoting community-based forest management recommend that certain types of people and certain types of land be excluded from the land market in order to prevent landholders from selling up and dispossessing themselves. In the Philippines, protection of this kind is built into legislation recognizing indigenous peoples' ancestral domains, which cannot be sold except by consent of all group members. The exclusion at work here is an exclusion from access to the land market. It also has a double edge: rural smallholders defined as indigenous often want fully individualized land ownership, and with it the right to take market risks in order to benefit from the wealth that is part—but only part—of the extraordinary growth that capitalism engenders. The extent to which activists, experts and authorities are currently devising schemes to divide citizenship, and peg types of landholding to types of person, is one of the more surprising findings that emerges from our comparative analysis, and an indicator that the dilemma presented by the market as a vehicle for exclusion is recognized by many actors—including some members of the designated groups—even if it is not yet effectively addressed.[15]

Claims to land on the basis of indigenous or ethno-territorial identity, in which one group asserts precedence and the right to exclude on the grounds of historical and affective claims to place, raise especially troubling dilemmas. Ethno-territorial claims are coming to the fore in many parts of Southeast Asia where decentralization programmes have reinvigorated customary and colonial concepts of ethnic belonging as the prime justification for land access, and transnational legal instruments protecting indigenous rights lend support. At the vicious extreme, exclusion has turned into violent eviction and ethnic cleansing. More often, the struggle has been peaceful but persistent, as "insiders" make it difficult for "outsiders" to acquire or hold on to land. National authorities are sometimes willing to concede ethno-territorial belonging as a legitimate basis for adjudicating claims. Yet the problem of who should have access is not solved. In much of Southeast Asia, decades of war, migration, eviction and compulsory resettlement make the question of who belongs where unusually fraught. Since it is not possible to send everyone back to their origins, the problem of how to allocate land to people who find themselves "out of place" but nevertheless assert a claim to life and livelihood still remains. Ethno-territorial claims also run up against other frames for allocating access (and legitimating exclusion) in the name of enhancing the welfare of the population, among them economic growth, conservation or the needs of the landless.

Finally, exclusion's double edge comes through again with respect to the characteristic of land administration in Southeast Asia that most frustrates experts, administrators and investors. Contradictory laws, inconsistent state agendas, overlapping allocations, shifting priorities, fuzzy boundaries, poor maps and incomplete data are usually interpreted as evidence of deficient state capacity, or a cover for land-grabbing. Yet this scenario has a productive dimension less often observed. Inconsistent laws enable different constituencies to argue that right is on their side. Laws recognizing the people's right to land coexist with laws justifying eviction in ways that legitimate one or another course of action, while leaving the underlying dilemma unresolved. Depending on the circumstances, guards charged with evicting farmers from a protected forest or unused land in the corner of a plantation can proceed with their task, with the law on their side, or they can choose to look the other way because they recognize that the farmers (who may include their own kin and co-villagers) are not just trying to survive—they are entitled. Fuzzy boundaries have the virtue of enabling flexible accommodations, and they are no doubt preferred by smallholders over the certainty of eviction should they find themselves on the wrong side of a definitive boundary line. Yet the lack of clarity has a significant downside: it enables officials at various levels to act as tyrants, using the power vested in them erratically to evict, intimidate, make a grab for resources or impose fines selectively.

In highlighting the double edge of exclusion as both a necessary element in the orderly and productive use of land, and a threat to lives and livelihoods, we have begun to expose the dilemma encountered by the authorities and experts who take responsibility for devising optimal arrangements, and by the NGOs and others who seek to challenge these arrangements on behalf of the rural poor. Nowhere in this book do we reduce the problem of access to a dichotomy in which exclusion (bad) can be counterposed to inclusion (good)—indeed, the very terms of such a dichotomy are incoherent, since the inclusion of some land uses, and some land users, necessarily means the exclusion of others. We have no prescription to offer, other than to suggest that it is better for those enacting policy, programmes and projects with excluding effects to acknowledge and understand the diverse implications of exclusion for different groups than to turn a blind eye to them. What we provide, in the pages that follow, is a careful exploration of *how* exclusion occurs. Our emphasis on the "how" question enables us to tease out the powers at work in different exclusionary regimes, and to place the question of who wins and who loses at the centre of our analysis.

In wrapping up this discussion of exclusion's double edge, we would like to comment briefly on the relationship between our notion of exclusion and the concepts of "enclosure", "primitive accumulation" and "accumulation by dispossession". During the 21st century, all three of these terms have been used widely in the literature on land relations in the global South. "Enclosure" is generally taken to mean the conversion of common property (most prominently land, but also other forms of commons such as knowledge and fisheries) into private property.[16] The concept of "primitive accumulation" has its origins in Marx. It refers to core elements of the process by which non-capitalist social formations are transformed into capitalist ones, in particular the separation of workers from direct access to the means of production, most notably through land enclosures that dispossess farmers and turn land into private property and capital.[17] David Harvey has recently re-conceptualized "primitive accumulation" as "accumulation by dispossession".[18] The terms themselves and the debates around them have been highly valuable, and we agree that all three of these processes are at work in contemporary Southeast Asia. Our use of "exclusion" is intentionally broader than any of these concepts, however, and thus encompasses a wider range of phenomena while also highlighting some lacunae and problematic assumptions in the use of these terms.[19]

Our main concerns about the literature involve assumptions about the "who" and the "why" of enclosure. Most authors in these debates assume, first, that enclosure/dispossession is undertaken by states and/or large corporations (or, anthropomorphically, by "capital"),[20] and, second, that it is undertaken for corporate profit or more broadly to facilitate accumulation. We demonstrate, however, that enclosure and dispossession are often motivated by quite different goals, such as the creation of conservation areas and the formation or protection of "ethno-territories".[21] We also highlight cases in which environmentalist NGOs, indigenous and other ethnic groups, village organizations engaged in community-based resource management, and smallholders acting on their own behalf have enclosed land and dispossessed others. A focus on this last group, in particular, brings out a third problematic assumption in the literature: that enclosure and primitive accumulation will be opposed by the poor. As our studies of cacao planting in the uplands of Sulawesi and the coffee boom in the Central Highlands of Vietnam show very clearly, smallholders in Southeast Asia have at times vigorously engaged in enclosure and primitive accumulation "from below". Tubtim and Hirsch also show how enclosure can be achieved, paradoxical as it may seem, through discourses of common property.[22] The often-intimate scale of what we might call these "micro-enclosures" should not blind us to the facts that smallholders do not always engage in community-oriented defence of the commons, and that they often *want* private property in land for themselves. That this desire can come freighted with the dilemma we discussed above does not make it any less real.

A fourth problematic assumption in the literature on enclosure is embedded in the way the process is envisaged as one in which states and corporations attempt to take away land that "communities" hold in common. While this is an appropriate description of some of the land being fought over in Southeast Asia, landholding in common is more the exception than the rule with respect to cultivated and residential land. In areas being "enclosed" for dams, land-titling programmes, peri-urban and tourism development, conservation areas, and plantations, much of the land is held, occupied and farmed privately by households. The fact that these holdings are often not recognized by the state does not make them "commons"; rather, they should be seen as individual property held under various informal and/or customary arrangements. These are arrangements that often permit the landholder to sell or mortgage the land, making it effectively a commodity, even though its transfer may be subject to some restrictions (see Chapter 2). None of this, again, is to deny that enclosure, primitive accumulation, and accumulation by dispossession, as defined by participants in these debates, are taking place in Southeast Asia, or to say that these concepts are not helpful in explaining why grazing lands, forests, wetlands and other "commons" are particularly susceptible to uncompensated expropriation. We also recognize that the notion of the commons has been used to frame a powerful defence of those facing dispossession in contexts that prioritize common over private rights. The discursive power of notions of "commons" and "community" to assert or defend land rights need not match the empirical status of "commons" on the ground. Nevertheless, the term "commons", like the term "enclosure", must be applied with caution and with attention to the potentially problematic assumptions embedded in it.

claims to land. Widespread military-backed land-grabbing in Cambodia is perhaps the most spectacular example in the region. Just as important, however, are more routine moments when the police act to defend the questionably acquired land access of powerful people. The range of actors we consider here thus includes state officials acting outside of their official responsibilities. It also covers the security guards hired to police haciendas in the Philippines and to prevent the theft of shrimp from ponds in Thailand, as well as the armed gangs that are indispensable to the power of the local godfathers and bosses we discuss in Chapter 5. Force, too, is not a monopoly of the powerful and well-connected; it is also used by the poor, and by smallholders, at a variety of scales. Smallholders may use arson against one another in village-level disputes and against state and corporate actors. Landless people occupy land by force; ethnic groups mobilize to intimidate each other, and indeed attack each other, over access to territory. In the Philippines the New People's Army, the armed wing of the Communist Party, is a key actor in disputes over land. Force is thus a ubiquitous underpinning of exclusion.

We stress, as well, that while outright violence is common, it is not necessarily the most important aspect of the way that force is used to exclude. Force can be extremely effective even when it is largely, or even entirely, implicit. People who are being relocated from the vicinity of a planned hydro dam might decide not to push for the compensation they have been promised because they feel that state actors would be prepared to use violence against their mobilization. People with a claim to land under agrarian reform in the Philippines may not fight for it because they know their landlord has access to extrajudicial force. State officials, too, worry about the potential for violence of their citizens, and undertake land reforms and other moves in hopes of pre-empting it. The possession of means of violence, then, can create a climate in which force acts quite effectively without ever being used. Such situations are linked to cases of actual violence through a continuum that runs through more or less explicit efforts at intimidation.

The third power that we consider in this volume is the *market*. Market forces have been of enormous importance to the dynamics of land access and exclusion in Southeast Asia. Most obviously, the price of land is a primary determinant of who can gain access to land and who cannot. As with the other powers we discuss, of course, defining what constitutes a market force is not straightforward. Markets do not spring into existence by themselves. They are underpinned, as mentioned above, by regulation, force and legitimation. States intervene constantly in markets as they try to shape economic activity and to reward favoured groups and clients. Land becomes available for sale only under certain conditions, and states try to suppress the existence of certain kinds of market (for instance, markets in land within conservation areas), with varying degrees of success. Crucially, too, participants in and regulators of markets are frequently the same individuals and groups. In our discussions of market powers, we try to keep these underpinnings and interventions very much in view. We also want to insist, however, that the *prices* of certain key commodities and services are critical to understanding the dynamics of exclusion. While these prices are not generated entirely in some abstract space of supply and demand, they do, for the most part, confront actors as hugely persuasive social facts. Rising world market prices for coffee have helped inspire millions of Southeast Asians to start planting the crop, whether on land they already held or on new land, a shift with profound exclusionary consequences. Spectacular rises in land prices on the rural-urban fringe and in tourist areas have encouraged smallholders to sell their land and given a spur to land-grabbing. The price of credit (or, put differently, the interest on debt) has been hugely significant in the processes of piecemeal land loss among smallholders we discuss in Chapter 6. And it is worth pointing out that there is often a "market price" for the bribe that will induce guards to look the other way as fishers use an ostensibly off-limits beach, or the pay-off that will convince a city council to rezone a piece of land for conversion.

The last of our four powers is *legitimation*. Understood as justifications of what is or of what should be and appeals to moral values, legitimations are of signal importance in supporting different forms of exclusion. Arguments framed in terms of what is right and appropriate provide the normative underpinning to regulatory, forceful and market powers, and many of the actors discussed in this book seek to reshape exclusion out of a genuine sense of moral duty. As we have already suggested, the range of legitimations for exclusion on offer in Southeast Asia is enormous,[25] and different discourses are often in direct conflict with one another. One family may feel that they have an unassailable right to a piece of land because they've paid good money for it, while another from a different ethnic group may see the same land as an inalienable part of their community's ancestral territory. Discourses of nation and citizenship may be mobilized by landless people to claim access to land at the same time that the goal of development—for the good of the nation—is being invoked to dispossess people to make way for dams and plantations. Claims to scientific knowledge, whether of the dynamics of a watershed or of the most efficient institutions of governance (including free land markets), can be used to trump conflicting claims and justify exclusion. More broadly, governments possessed of the "will to improve" claim to know how people should live and to have both the right and the duty to get them to live that way.[26] While legitimations are powerful, then, they are never unopposed, and the effort to justify any particular form of exclusion must always be seen as a struggle involving a wide range of actors. Increasingly, too, these struggles take place at a variety of scales, and actors invoke different conceptions of scale—from the local to the global—in the arguments they make. As Ribot and Peluso note, for example, appeals to the notion of a "global commons" by international NGOs and others "create universalizing categories and naturalize their interventions around the world in the name of environmental protection".[27]

Regulation, force, the market and legitimation, then, are the four powers that actors seeking to exclude must deploy, or to which they must respond. They are variably effective across different scales. Force operates most effectively when it is employed at a proximate scale (when one party has the capacity to inflict physical harm on another), but as we noted earlier it can also operate at a distance, in the form of a pervasive threat. Legitimation is always a matter of appeal to an audience, and it matters greatly whether the two parties share a discursive frame (e.g., two sisters disputing their inheritance) or are separated by physical and social distance (farmers here, a transnational lobby promoting conservation-by-eviction there). A receipt for land taxes, issued by a local official, may be scaled up by landholders as evidence that "the state" has acknowledged the legitimacy of their claims. The power of the market operates most concretely at close quarters—"This is the price the land broker is willing to pay us for our land, today." But in its abstract form, as a hope or promise, it is still powerful enough to provoke a response. The most spectacular example is when visions of riches prompt smallholders (who do not have access to market research) to rush into the production of the latest boom crop. The complexity of the narratives that we introduce in the empirical chapters thus derives not only from the range of actors and powers involved but also from the different ways in which powers are mobilized across scales.[28]

We end this discussion of our four powers of exclusion by observing that none of these powers operates in a frictionless world. Implementing effective rules, buying or selling newly valuable land, threatening tenants, convincing people that one's view of the world is the right one: all of these actions entail costs in terms of money, time and effort, and can easily go awry. Authorities cannot evict all the people all the time. Eviction is both expensive and disruptive, as the losers do not quietly disappear or starve. They protest, they disrupt business, they occupy land, and they reassert their right to survive in ways that even the most authoritarian regimes cannot entirely ignore. Thus, active compromise or, more often, a vague and inarticulate inertia in which things are left as they are, are constant features of exclusion. Like the inconsistent legal and policy regime we described earlier, they can be seen as ragged but enduring responses to the dilemma that exclusion presents.

Unfolding Powers of Exclusion

In the chapters to follow, we explore six key processes that are driving changes in land access and exclusion for large segments of Southeast Asia's rural population, through deployment of the powers we have identified. In each chapter, we combine broadly comparative, conceptual discussions with detailed studies of selected times and places where the dynamics of exclusion have been strikingly revealed. The focus of Chapter 2 is *land formalization and allocation,* a process expanding rapidly across the region and one driven mainly by state agendas to regularize land access, promote development and consolidate territorial control. In some cases, state projects formalize land ownership, and hence exclusionary rights, in support of market-based development and markets in land. Land titling epitomizes this process, and we examine the way this has played out in Thailand and Laos. Elsewhere, formalization has been partial or hybrid in character, deliberately setting aside areas in which fully alienable title is not to be granted and instead allocating land with restrictions based on zoning and limits on transferability. We look at the Land and Forest Allocation programme in Laos to illustrate this. States have also been complicit in nominally redistributive processes, notably land reform, although this has tended towards the formalization of rights to public land rather than large-scale progressive transfers, as seen in the case of the Philippines.

In Chapter 3 we explore the territorial expansion and intensification of *conservation,* a process in which state objectives are combined with transnational donor and NGO concerns to regulate peoples' impact on the environment, broadly conceived. Exclusions legitimated by environmental concerns have become ubiquitous—as ambient as the air we breathe, and often unremarked. They are supported by notions of the "common good" but have quite uneven social effects. We review the most familiar form of conservation-based exclusion, the establishment of protected areas that forbid human use, drawing an example from Sulawesi. We then explore a less-studied but increasingly significant form of exclusion that is dispersed throughout the countryside, as villagers are invited by NGOs and other agencies to participate in establishing a degree of self-exclusion under the label "community-based natural resource management". We illustrate this with examples from Cambodia. Our third focus is the practice of setting aside land and other resources for protection to mitigate the environmental impacts of large-scale resource projects. This practice creates a kind of double exclusion, as people are displaced or otherwise restricted in their access to land by both the project *and* its mitigation efforts. A large dam in Laos provides our case in point here.

Chapter 4 examines the massive rearrangements of land access and exclusion associated with the occurrence of crop booms, which are a long-standing feature of agrarian Southeast Asia. Crop booms are stimulated by rising commodity prices, the introduction of new growing techniques, and policy interventions, in varying configurations that we explore through a focus on three of the most important contemporary boom crops: oil palm (with a focus on Sarawak, Malaysia), farmed shrimp (especially in Thailand) and coffee (which boomed spectacularly in the Central Highlands of Vietnam in the mid-1990s to late 1990s). Some of the most significant similarities between these booms have been large flows of migration to boom areas, a strong tendency for people to make more individualized claims to land, land-grabbing, deforestation and the combination of genuine gains (including for smallholders) with the risk of immiseration.

Chapter 5 again highlights the power of price, but in this case the market is pushing people out of agriculture rather than towards it. As the industrial and service sectors boomed across parts of Southeast Asia in the 1980s, 1990s and 2000s, post-agrarian land uses began to compete sharply with agriculture. One aspect of this transition is the sprawl of cities and the creation of distinctive peri-urban zones that mix industrial, commercial, infrastructural, residential and agricultural land uses. Such zones have been growing at extraordinary speed around many of Southeast Asia's cities, and their development (as illustrated by our focus on Cavite, a province near Manila) has been profoundly shaped by the interaction of our four powers of exclusion. Tourism is a second land use that has prompted the conversion of large amounts of agricultural land, through processes that are often quite similar to those that characterize peri-urbanization. We show this through studies of tourism in Bali and Angkor. We also take up the quite different exclusionary dynamics that characterize the construction of hydroelectric dams, with a focus on the Hoa Binh Dam in northern Vietnam.

In Chapter 6 we examine the processes through which social intimates, neighbours and kin, exclude each other from access to agricultural land. This process is long-standing in Southeast Asia, especially in the more densely settled agricultural cores. Yet it merits renewed attention as competition among smallholders intensifies and expands to frontier areas (driven sometimes by crop booms) at the same time as land frontiers are closing, making it increasingly difficult for people who lose their land to move off to try again in another location. We examine "intimate exclusion" in village Java, the site of classic debates about whether villagers "share poverty" by spreading access to land and work or, alternatively, exclude intimates from access. Our second study describes a process of "enclosure from below" followed by a rapid process of agrarian differentiation among highland farmers in Sulawesi who previously shared access to common and collectively inherited land. Finally, we turn to

Vietnam, to examine the re-emergence of unequal land access among villagers and the debates that have emerged among villagers over the rights and wrongs of exclusion.

In Chapter 7 we examine *counter-exclusions* in which groups mobilize to reclaim land from state authorities, or from other groups in rural society whose presence they deem illegitimate according to criteria such as social justice or ethnic belonging. A striking feature of these mobilizations, in contrast to the class-based mobilizations of an earlier era, is the extent to which they draw legitimation from state failure to deliver on the promises of even-handed citizenship enshrined in national constitutions, or from notions that link development to improvements in popular well-being. These features are prominent even in egregious cases such as the violent expulsion of migrants from ethno-territorial homelands in Indonesia, our first empirical study, and in the more peaceful attempt by indigenous groups in Vietnam's Central Highlands to restore their autonomy in the face of a massive migrant influx. They also arise in mobilizations against eviction, the topic of our final section, which takes up Thailand's Assembly of the Poor. Taken together, these mobilizations starkly reveal exclusion's double edge. A state that fully recognizes ethno-territorial rights cannot simultaneously treat land as a national resource to be distributed to the landless on the basis of need; nor can protests against eviction be easily reconciled with demands for development when exclusion is necessary for productive investment. It is often easier to discern who benefits and who loses than it is to resolve the dilemma at the heart of exclusion from land.

As we proceed into our empirical analysis, we want to remind readers that Southeast Asia is a region of extraordinary diversity by almost any metric that one might care to use. While we believe the comparative and synthetic analysis we offer here is appropriate to the subject matter, we recognize the risk that a reader could get lost. To alleviate this problem, we have prepared maps that show the main features of regional topography and the places named in the text, and some tables that summarize demographic statistics. Throughout, we have attempted to write in a way that will be insightful for those familiar with the region, yet accessible to readers new to Southeast Asia. For those who would like more background on the countries covered in the book (Cambodia, Indonesia, Laos, Malaysia, the Philippines, Thailand and Vietnam), we have prepared an appendix that contains brief histories of national land relations. We also explain some key terms in a short introduction to the historical geography of agrarian Southeast Asia in the first section of the following chapter.

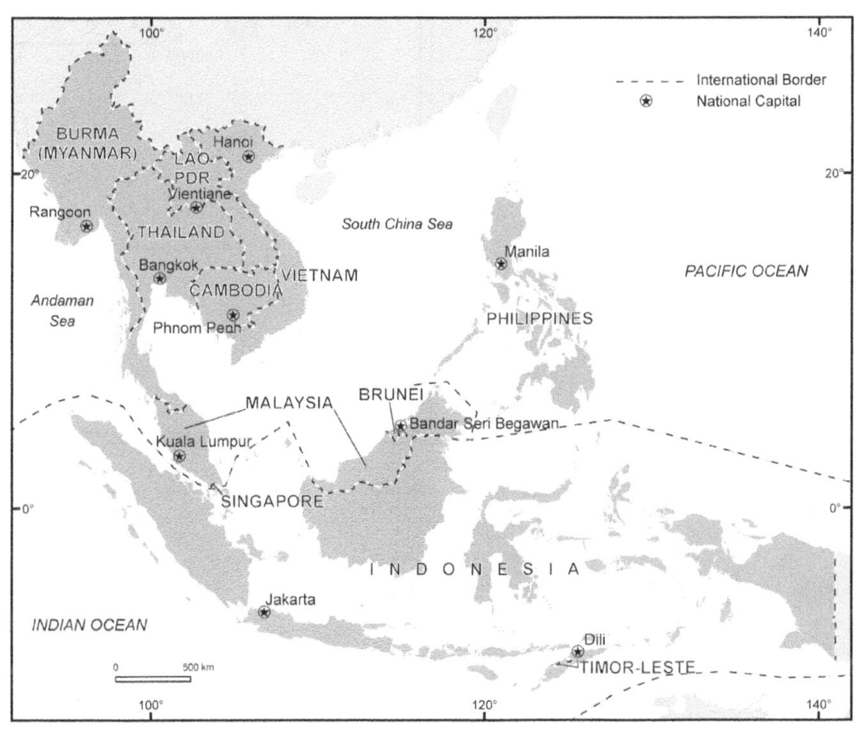

Figure 1.1 Southeast Asian countries and capital cities

Sources: Digital Chart of the World, Environmental Systems Research Institute.

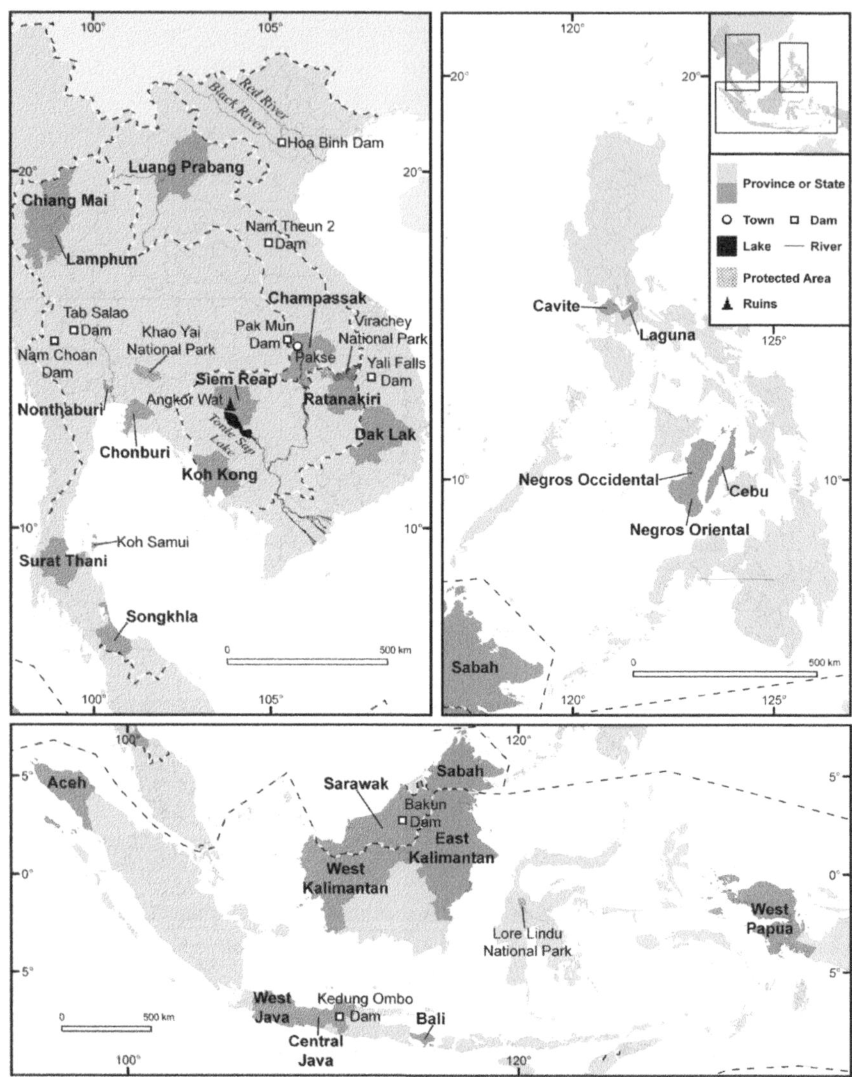

Figure 1.2a Southeast Asia: Provinces, dams, rivers and other places mentioned in the text

Sources: Digital Chart of the World, Badan Nasional Penanggulangan Bencana (Indonesia), Vietnam Administrative Atlas, Malaysia Map and Guide, Department of Local Administration (Thailand), Department of Geography of the University of the Philippines Diliman, Central Statistical Organisation (Burma), National Committee for Sub-National Democratic Development, Lao Department of Statistics, Atlas de Timor Leste (Universidade Técnica de Lisboa), Food and Agriculture Organization (United Nations), Landsat, World Database on Protected Areas.

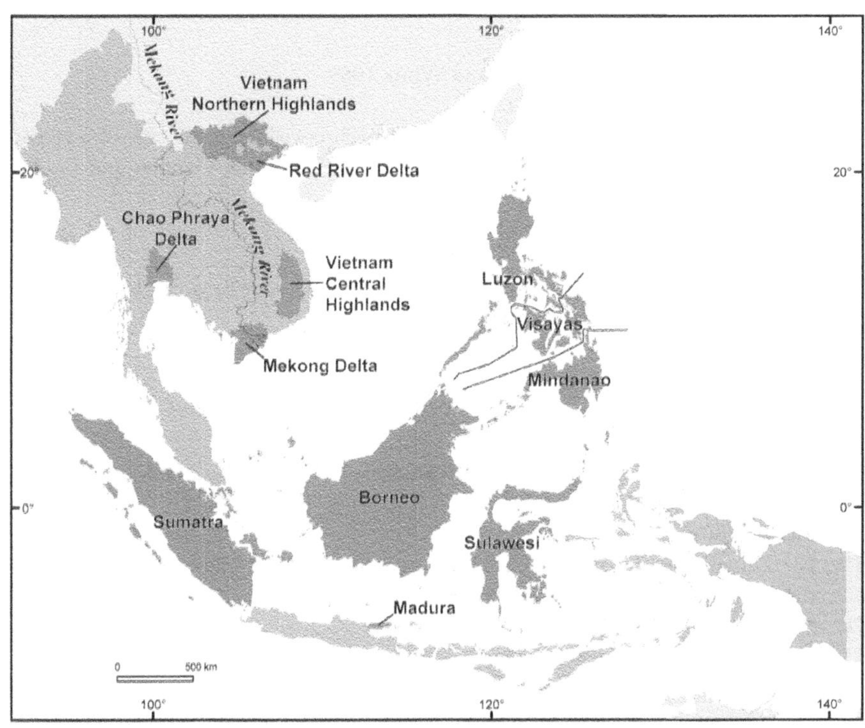

Figure 1.2b Southeast Asia: Regions mentioned in the text

Sources: Digital Chart of the World, Badan Nasional Penanggulangan Bencana (Indonesia), Vietnam Administrative Atlas, Malaysia Map and Guide, Department of Local Administration (Thailand), Department of Geography of the University of the Philippines Diliman, Food and Agriculture Organization (United Nations).

Notes

1. See Appendix, and the database and maps prepared by the Cambodian Human Rights portal, http://www.sithi.org/temp.php?url=landissue/mapping_dev.php and http://www.sithi.org/landissue/db/All%20Concession%20List.pdf.
2. *Phnom Penh Post*, 28 Sept. 2009.
3. Seng 2004.
4. See Fox 2002, Chhay 2004.
5. For an example, see Hart, Turton and White 1989.
6. Barkin and Shambaugh 1999: 4.
7. Borras and Franco 2010: 19. A similar approach is visible in the title of a recent report from the Rights and Resources Initiative: Sunderlin, Hatcher and Liddle, "From Exclusion to Ownership?" 2008.
8. Sikor and Thanh 2007: 650, 652.
9. Ribot and Peluso 2003: 153.
10. Ribot and Peluso 2003: 156, emphasis in original; see more broadly 155–61.
11. Li 2010b.
12. Polanyi 1957.
13. See, for instance, Wood 2002.
14. Li 2009a.
15. Li 2010b.
16. See *Ecologist* 1993; De Angelis 2004; Vasudevan, McFarlane and Jeffrey 2008.
17. Perelman 2000, Glassman 2006.
18. Harvey 2003.
19. For a discussion of primitive accumulation as an analytical framework for studying rural Southeast Asia, see Hall 2010.
20. See De Angelis 2004.
21. Michael Webber has made a very similar argument about rural China, writing that "primitive accumulation depends not only on the logic of competition between capitalist and non-capitalist forms of production, but also on extraeconomic agendas—in this case, of environmentalism and ethnicity" (Webber 2008).
22. Tubtim and Hirsch 2005.
23. Verdery 2003.
24. For example, while our typology resembles the division of power into ideological, economic, military and political forms that Michael Mann develops in *The Sources of Social Power*, we use our terms in a less all-encompassing way than does Mann (1986). For another very similar typology, see Vahari 2009.
25. See also Bakker, Nooteboom and Rutten 2010.
26. Li 2007c.
27. Ribot and Peluso 2003.
28. See Allen 2003 for an illuminating discussion of the way different powers work across scales.

2

Licensed Exclusions: Land Titling, Reform and Allocation

Throughout Southeast Asia, land access is being reconfigured by state programmes that seek to include small farmers in formalized access to agricultural land, or transfer land ownership to them. These programmes license the access of one group of actors, an intervention that simultaneously, and necessarily, licenses the exclusion of another. For the most part, these programmes address farmer demands. Farmers need access to land, and to use land productively they need to exclude others from it. Given the option, they usually want to have secure title. If titling is not available, they still welcome other programmes that formalize their land occupation and use. Landless farmers often press states to undertake land reforms and are willing to move to areas where governments allocate land under settlement programmes.

Inevitably, the exclusions produced by licensing present dilemmas. For land policy makers, programmes that intervene in land access require the selection of winners and losers. NGOs and activists sometimes find themselves campaigning against the issuing of property rights that the people they claim to represent actually want. And while few farmers are inclined to turn down state-recognized rights to land that they occupy, they find themselves chafing against the new regulations, costs and conditions of use that accompany such recognition. Often, they discover a gap between what land formalization schemes promise and what they can deliver.

State projects designed to improve access to land for smallholders and landless farmers can be divided into four types. The main goal of the first two—formalization and titling—is to give farmers a more secure, state-recognized claim to land that they already hold. Land *formalization* refers to the recognition and inscription by the state of rights and conditions of access within specific boundaries. Land *titling* is a specific type of formalization, one

27

in which the state demarcates the boundaries of the land, records ownership in a cadastral registry, and recognizes the landholder's ownership of the land and the right to sell, transfer or mortgage it. States recognize titled land as a saleable commodity, and full title is often seen as the evolutionary "end point" of formalization. The second two types of project, land reform and land settlement, also involve formalization in that states generally give beneficiaries some kind of formal claim to their land. What makes them distinct is that rather than aiming to formalize access for people already in possession of the land, they seek to transfer land to them. In *land reform* projects, currently owned land is redistributed to the landless or to smallholders with very little land. *Land settlement* programmes support the movement of people to areas with land (generally public land) that is ostensibly unused and available for farming. While the allocated land is very often already claimed and used by people in the area, land settlement differs conceptually from land reform in that it claims to be opening up "new" land on the "frontier" rather than redistributing already-owned land. Variants on land settlement include the resettlement of farmers displaced by development schemes, and consolidation projects in which state authorities try to move frontier villages and smaller scattered communities into larger, more accessible clusters.

To understand why Southeast Asian states have been so concerned with formalization, titling, land reform and resettlement, we need some background on the historical geography of the region's agriculture. This geography is often understood in terms of a distinction between agricultural cores and frontiers, a binary conceptualization that is often misleading but nevertheless retains value as a point of orientation. Southeast Asia's agricultural cores are located largely in river deltas, valleys and plains. They are characterized by dense smallholder populations, usually members of the ethnic majority of the nations in which they are located, who grow food crops (mainly rice) under a wide range of tenure conditions. Frontiers have historically been less densely populated than cores and more oriented to both swidden and non-food-crop production. Their inhabitants have generally been viewed by elites and sometimes "core area" farmers as somewhat untrustworthy: people marked by ethnic and/or cultural difference, backwardness and questionable attachments to national identity who pose threats to national security.

The current division between cores and frontiers in Southeast Asia has come into being over the last 200 years. Most of Southeast Asia was very sparsely populated before the mid-19th century.[1] Since then, both population and cultivated area have expanded dramatically as settled agriculture has spread out from the river valleys and coasts to colonize swamps and forestland. Across much of the region there were two major spurs to agricultural expansion. One began in the middle of the 19th century, when improvements in shipping and changes in trade policies deepened Southeast Asian integration into world markets and sparked the opening of massive amounts of land.[2] Many of today's agricultural cores, especially the great rice-growing river deltas, were opened for cultivation at this time. The other major shift began in the 1950s, when the region's non-Communist states saw a push for expansion, primarily into the uplands, on the back of new crops and new technologies, strong world market demand, road building and other state support, and rapid population growth. Much post-World War II agricultural expansion has involved the production of cash crops for export. In some areas (notably Thailand), this has taken place through smallholder colonization of the frontier; in others (especially the insular states of Indonesia, Malaysia and the Philippines), plantations and estates have played a greater role. In the Indochinese states, expansionary dynamics were blunted at first by Communist governments and warfare, but later actively promoted by centrally planned land colonization schemes.

Across Southeast Asia, then, the movement of smallholders to frontier areas has been fundamental to agricultural expansion. Often, people have lit out for the frontier to flee oppressive conditions and in the hope of making new lives for themselves far from the reach of the state. States, however, have often encouraged people to make this move. Governments have supported agricultural settlement in many ways, from building infrastructure (canals, railways, roads) to organizing and subsidizing formal resettlement programmes. They have also tried to tame, to channel and, at times, to prevent such movements. A rough sense of the demographic rationale for population movements can be gleaned from Figures 2.1 and 2.2, which together indicate the concentration of population in the lowlands and sparse settlement in highland areas, and Table 2.1, which gives a snapshot of population densities in each nation.

The conflicting goals of state actors attempting to channel population movements will be a common theme throughout this and subsequent chapters, but for present purposes two issues are critical. First, agricultural expansion during the 20th century far outstripped the state's ability to keep track of, let alone effectively regulate, who had what land and what they were doing with it. Land formalization has thus, to a substantial extent, been a story of states playing catch-up. Second, since the colonial period, but more systematically since World War II, forestry departments and other state agencies have tried to keep large areas of land off limits to agriculture, primarily to conserve forests. We discuss the "political forest", or land declared by states to be forest, in more detail in Chapter 3, but it is vital to this chapter because its awkward relation to agricultural expansion has left tens of millions of Southeast Asians practising agriculture on land that is legally off limits to farming. Its extent is indicated in Table 2.2. Across the region, too, these people and this land have been subject to a fragmented and overlapping division of governmental responsibility in which multiple agencies jealously compete for authority over land. This contested jurisdiction may, for instance, see farmers paying agricultural land tax to one state agency on land that another says may not be farmed. While "the state" is at the heart of this chapter, then, "it" is not a unitary actor pursuing consistent and coherent goals. Rather, the motivations of state actors have frequently been not just multiple but contradictory.

The projects we discuss in this chapter are also linked historically by the ways in which states have used them sequentially in pursuit of the same objectives. Titling, formalization, land reform and land settlement are usually treated as separate—even competing—elements of policy and practice. We argue, in contrast, that these programmes share a common genealogy in state

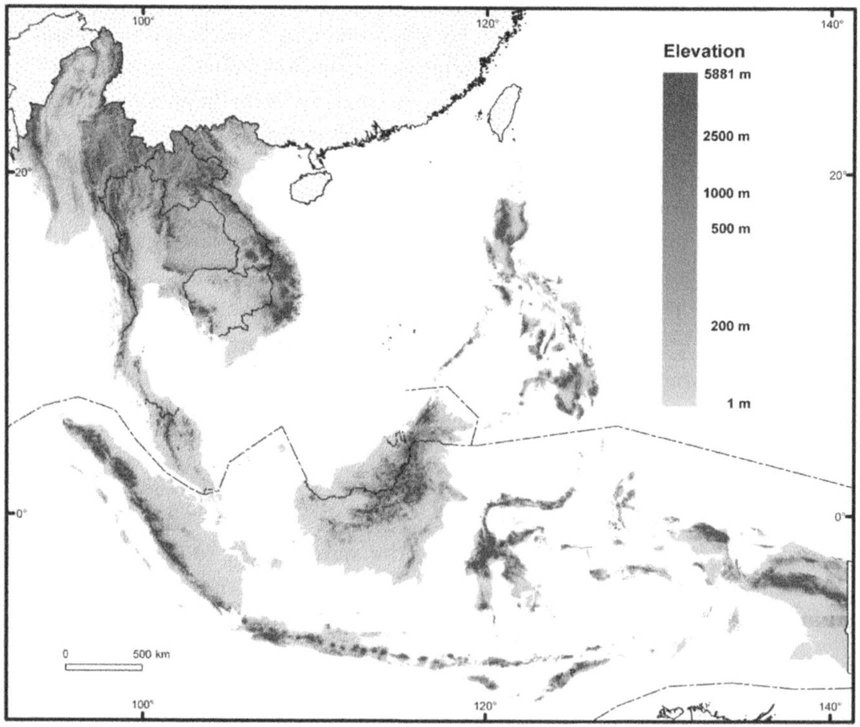

Figure 2.1 Southeast Asia: Elevation above sea level and topography

Sources: Digital Chart of the World, Consultative Group for International Agriculture Research - Consortium for Spatial Information (CGIAR-CSI).

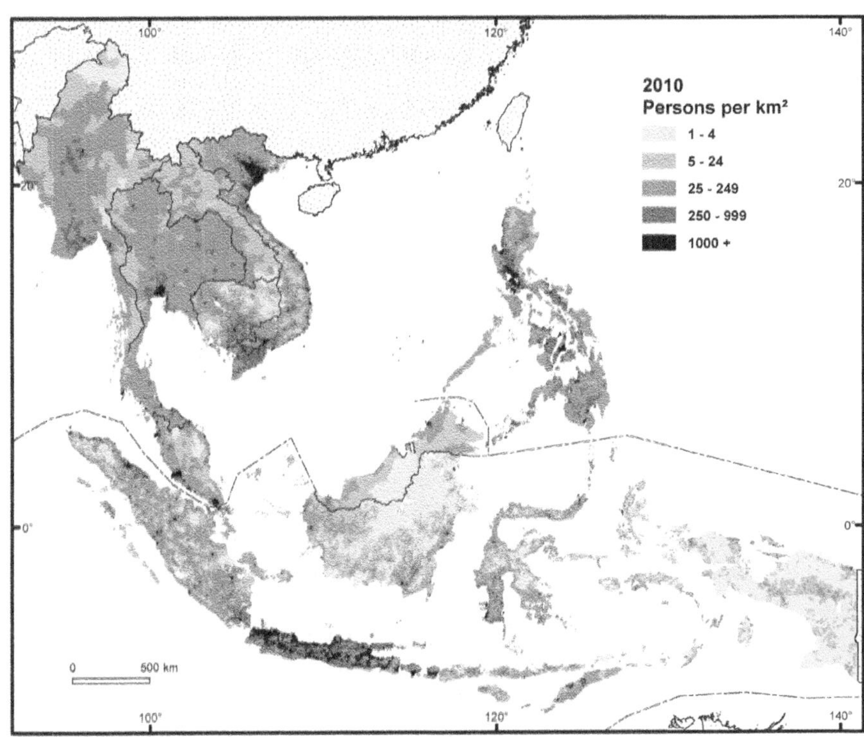

Figure 2.2 Southeast Asia: Population density, 2010

Sources: Digital Chart of the World, Center for International Earth Science Information Network (CIESIN, Columbia University), Centro Internacional de Agricultura Tropical (CIAT).

Table 2.1 Southeast Asia: Population density per square kilometre, 1980 and 2005

	1980	2005
Cambodia	37	77
Indonesia	77	115
Laos	14	25
Malaysia	42	78
Philippines	160	285
Thailand	92	129
Vietnam	161	253

Source: United Nations, http://data.un.org/Data.aspx?q=density&d=PopDiv&f=variableID%3a14.

 Land Dilemmas in Southeast Asia

Table 2.2 Political, biological and conservation forests in Southeast Asia as percentage of national land, 1970s and 2000s

	Biological forest (%)		Political forest[1] (%)		Conservation/ Protected area (%)	
	1970s	*2000s*	*1970s*	*2000s*	*1970s*	*2000s*
Vietnam	34[2]	31[3]	*	58[4] (1997)	3[5] (1980)	12[5]
Cambodia	73[6]	59[7]	*	?[8]	0[5] (1980)	25[5]
Laos	54[9]	42[10]	*	62[11]	1[5] (1980)	24[5]
Thailand	39[12]	31[13]	49[14]	46[15]	5[16]	25[4]
Malaysia	**	58.5[23]	**	53[18]	**	4.1[23]
Sarawak	55[17]	45[24]	72[24]	64[24]	0.6[19] (1981)	8[20] (2010)
Indonesia	70[21]	51[22]	70[23]	57[24]	26[25]	27[26]
Philippines	34[27]	19.3[25]	**	45.19[28]	1.0[29]	11.8[26]

Notes: The diversity of sources for this data means that caution is needed in comparing one country to another, since the criteria for the figures vary significantly.

* It is not meaningful to talk of "political forest" in Vietnam, Cambodia or Laos in 1975, as all three were in a state of immediate post-revolutionary regime change, without a forestry administration asserting territorial jurisdiction over land.

** Data not available.

[1] Land area under the control of national forest departments, not all of it with biological forest cover. The figure includes conservation areas.

[2] Figure for 1976, natural forest only. World Bank, Ministry of Natural Resources and the Environment; Danish International Development Agency, *Vietnam Environment Monitor 2002.*

[3] Figure for 2004, natural forest only. World Bank, Ministry of Natural Resources and the Environment; Swedish International Development Agency, 2005; *Vietnam Environment Monitor 2005.*

[4] Nguyen 1997, p. 9.

[5] International Centre for Environmental Management 2003, p. 21.

[6] Figure for 1965. Technical Working Group on Forestry & Environment 2007.

[7] Figure for 2006. Technical Working Group on Forestry & Environment 2007.

[8] It is premature to talk of an area of "political forest" in Cambodia, since the National Forest Program there has not demarcated forest areas in the same way as it has been done in neighbouring countries. The Forestry Administration administers logging concessions, the Ministry of Environment administers protected areas, and the Fisheries Administration administers mangrove forests and flooded forests around the Tonle Sap Lake.

9   Figure for 1973. Domoto 1997.

10   Figure for 2002. Department of Forestry, 2005.

11   Department of Forestry, 2007, based on total of conservation forests, protection forests and production forests.

12   Royal Forest Department.

13   Royal Forest Department.

14   Feder, Onchan and Chalamwong 1988.

15   Figure for 1997, based on a report that 10.1 million rai (16,160 square kilometres) of public land, or about 3 per cent of the national land area, had been allocated for Agricultural Land Reform and hence taken out of the political forest. Chirapanda 1998.

16   ICEM (2003) reported with reference to the Third NESDP (1972–76) that the Thai government had declared 9 national parks, 11 wildlife sanctuaries, 9 non-hunting areas and 6 forest parks, covering 4.9 per cent of the country.

17   Jomo *et al.* 2004. Political forest here is taken as Permanent Forest Estate.

18   Woon and Norini 2002.

19   State Planning Unit, Sarawak, c. 1983. Figures kindly provided by Rob Cramb.

20   Source: http://www.forestry.sarawak.gov.my/forweb/sfm/tpa1.htm.

21   The 1950 forest cover including plantations was 84 per cent. By 1985 the figure was 63 per cent (FWI/GRW 2002: 8, 12). Since deforestation accelerated after the Forest Act of 1967, we have estimated forest cover in 1975 at 70 per cent.

22   Actual forest cover is for 1997 (FWI/GRW 2002: 81) and does not reflect extensive logging after Suharto resigned and decentralization measures were implemented. See McCarthy 2000.

23   Critics argue that the Forest Department failed to follow its own procedures for gazetting the state-claimed forest estate, making its claims effectively "extra-legal". See Fay and Sirait 2002.

24   Official data from the Department of Forestry, http://www.dephut.go.id/files/IV1102.pdf, accessed 7 July 2009.

25   Figures for 1986; conservation forest 10 per cent of national land area; protection forest 16 per cent (FWI/GRW 2002: 13, 16).

26   Official data from the Department of Forestry, http://www.dephut.go.id/files/IV1102.pdf, accessed 7 July 2009. Conservation forest here includes parks and nature reserves (12 per cent) and protected forest (15 per cent).

27   Figure for 1970. "Decline of the Philippine Forest", Environmental Science for Social Change, Booklet with accompanying maps. Quezon City: Ateneo de Manila University, 1999, p. 15.

28   Wenk forthcoming: 68. This figure of area classified as forestland plus area classified as protected areas is for the early 2000s (from a 2004 source, no indication of date).

29   Source: Wikipedia list of national parks in the Philippines, http://en.wikipedia.org/wiki/List_of_national_parks_of_the_Philippines.

concerns with *legibility* and *agrarian unrest*. Government actors have used them to make the conditions of land use and access more "legible", that is, more comprehensible to and governable by the state.[3] This task is carried out by means of technologies of surveillance that include maps, surveys, cadastral registers, fences and boundary posts. Formalization projects are motivated in substantial part by official desires for better information about who is using what land for what purposes and for the revenue that comes from taxing newly "visible" resources,[4] while resettlement programmes promise expanded state control over sparsely populated "frontier" areas.[5] It is not only governments that are keen on legibility. The increased legibility brought about by both the Land and Forest Allocation programme in Laos and Certificates of Ancestral Domain in the Philippines (discussed below) has been welcomed by companies looking to invest in plantations.[6] Perhaps more surprisingly, farmers on state land often go out of their way to be seen to pay land taxes in order to secure tax receipts that serve as legible evidence, and tacit state recognition, of their right to farm in the political forest.[7]

State actors have also used these programmes to respond to and pre-empt agrarian unrest. Since World War II, all Southeast Asian states have seen serious rural political mobilization calling for more equal access to agricultural land (see also Chapter 7). While redistributive land reform is the obvious political response to such mobilization, it has come up against the implacable opposition of landowners in all but the most revolutionary situations. Hence, states have turned to the frontier as a terrain where land can be found without taking on powerful actors, and thus as a "safety valve" for people living in crowded core areas who are mobilizing, or might mobilize, against the government.[8] In other circumstances, land reform has given way to titling, based on the neo-liberal principle of solving small farmers' problems by making them property owners.

This chapter explores the exclusions and powers at work in land titling, formalization, reform and settlement schemes, drawing on examples from across the region. In the first section we take up the World Bank/AusAID-funded titling programmes and compare titling in Thailand and Laos. We then examine land formalization schemes (mostly on public land) that fall short of full titling. We argue that state and other actors are pushing for a surprising amount of Southeast Asian land to be protected from the full force of market power, a point we illustrate with reference to the Land and Forest Allocation programme in Laos. We then consider land reform and land settlement projects, and illustrate the slide from redistributive land reform towards land settlement and formalization of settlement on public lands, through a study of the Philippines' Comprehensive Agrarian Reform Program.

En-titling Exclusions

We begin our study at what might be called (with a heavy dose of irony) the "end of history". Land titling evokes this term both because it is often seen as the natural end point of land rights formalization—the point at which the process is "finished"[9]—and because it attempts to bring about the individualized neo-liberal model that Francis Fukuyama presents as the end of contestation over the appropriate economic organization of society.[10] By the early 21st century, large-scale land titling projects aimed at ending land history had become core pillars of rural development policy in Thailand, Laos, the Philippines and Indonesia. These were not the first efforts to title rural land in Southeast Asia. Many colonial regimes sought to attract investment in plantations by creating administrative systems that would assure European investors of the security of their claim to land.[11] Some colonial governments were convinced that individual title would improve the welfare of native populations. During the early 20th century, both the British in Malaya and the Americans in the Philippines attempted projects of this sort, though with limited success.[12] In uncolonized Siam, too, land in some of the rice-growing areas of the Chao Phraya Delta was titled. In all cases, however, the amount of titled land remained small as a percentage of agricultural land (to say nothing of national territory), and titling was given low priority as a national project until the 1980s.

Land titling in many states across the region has been linked by a common approach. Titling projects in Thailand, Laos, Indonesia and the Philippines are based on a collaborative arrangement between the World Bank, Australia's bilateral development assistance agency AusAID, and a private managing contractor, Land Equity International (LEI). The model was first implemented in Thailand, and the Thai programme (the largest in the world) has come to serve as a model of World Bank best practice not only in Southeast Asia but throughout the South.[13] The model sees the World Bank providing loans for broad-scale implementation, while AusAID gives grants for technical assistance implemented by LEI. The latter works with the relevant national agency and adapts its approach to each country's land law. These arrangements have resulted in a significant continuity in the personnel working on titling projects across the region, whether they work for LEI or the World Bank.[14] The titling model itself is based on Australia's Torrens land registration system, which has also served loosely as the basis for Thailand's land registration system since 1901.[15]

The double edge of exclusion that we highlight throughout this book is nowhere more evident than in land titling. Titling is the most comprehensive manifestation of the modernist project of clarifying and consolidating access to land, and it represents to its proponents the perfection of exclusion. By definition, land title is the right to exclusive use of a parcel of land, irrespective of the social identity of the owner—unlike customary tenure systems, in which the identity of the landholder is often key. The holder of a title has the right to exclude any other party from using and accessing land and has the exclusive and complete right to dispose of land through sale or mortgage. A title represents a complete "bundle of rights", including the right to exclude other parties from benefits such as seasonal use, physical access and right of way across the land. This perfection of exclusion is not just the definition of title, it is also the goal that titling programmes aim to achieve by removing what are seen as the insecurities and ambiguities of customary tenure systems.

Land titling proponents, however, describe titling primarily in terms of *inclusion*. Assigning full property rights in land, they claim, is akin to including small farmers *qua* landowners as legal and market subjects, fully fledged players in an integrated market economy. The Peruvian economist Hernando de Soto, one of the most prominent advocates of land titling, argues that poor people without full title to their land are excluded from the modern economy by a metaphorical "bell jar" and that titling is the way to let them in.[16] For de Soto, the "property effects" associated with full land title are the most powerful mechanism available for pulling the poor out of poverty, and for promoting development more generally.[17] As smallholders with title become wealthier through property ownership, access to institutional credit and enhanced confidence to invest in land-based production, and as newly lubricated land markets transfer land to those able to use it most efficiently, growth will take off. The economist Helen Hughes has gone so far as to assert that without private, individual land tenure, development is impossible.[18] Southeast Asian leaders such as Gloria Macapagal-Arroyo (former president of the Philippines), Susilo Bambang Yudhoyono (president of Indonesia) and Thaksin Shinawatra (former prime minister of Thailand) have expressed enthusiasm for de Soto's ideas.[19] The World Bank has further expanded on the virtues of titling by suggesting the following:

> [O]ver and above the economic benefits that may be derived from giving households greater tenure security, measures to increase households' and individuals' ability to control land will have a clear impact on empowering them, giving them a greater voice, and creating the basis for more democratic governance and participatory local development.[20]

This, then, is the democratic and populist promise of titling: to include the poor within the national economy, national politics and the national legal system. In the next section, we consider how titling works out in practice.

Titling in Thailand and Laos[21]

If the extension of title to all of a country's agricultural land marks "the end of land history", some countries in Southeast Asia are closer to this putative end point than others. Land titling programmes in Thailand (which commenced in 1984 with a 20-year time horizon) and Laos (which started 13 years later, in 1997) have been implemented against significantly different histories of land administration. In Thailand, titling has followed a long, slow process of land formalization in more accessible areas. The 1954 Land Code established a progression of security of title on private land from reservation certificates (NS2), which expire if land is left unused, through utilization certificates (NS3), to full land titles or *chanood* (NS4). This graduated system was instituted to discourage speculative land-grabbing on the frontier, and to encourage the people claiming land to use it productively. The certificates and titles are not available on land defined as state land, which is under the control of the Royal Forest Department or other state agencies. During the 1960s and 1970s, rapid expansion of agricultural land outstripped the capacity of the Department of Lands to carry out land surveys and issue certificates.[22] By the early 1980s, only 12 per cent of the 23.7 million hectares of land under cultivation had been fully titled. A huge amount—39 per cent—had no land certification whatsoever, or was illegally occupied forest reserve land, while the remainder had some form of official recognition short of title. The titling project was designed to apply to land plots where there were already NS2 or NS3 certificates.

The Land Titling Project was established in 1984 to support the Department of Lands (DOL).[23] Part of its rationale was simply that it would take more than 200 years to achieve full coverage of non-state lands at the prevailing rates of registration. Proponents of land titling in Thailand claimed that it would also have other benefits, including benefits framed in environmental terms. At the time, there was growing concern about illegal agricultural encroachment on rapidly dwindling forests. Titling, the DOL argued, would help solve this problem, as enhanced tenure security and access to collateral would encourage farmers to invest in intensified production on their existing land. They proposed that consolidating exclusion at the level of individual farm plots should proceed together with the consolidation of exclusion from forests, where boundaries should be clearly demarcated.

The titling project was highly successful on its own terms. By 1998 it had increased the number of title deeds on the country's 26 million parcels from 4 million to about 19 million.[24] Land increased in value, there were more land transactions, and there was increased borrowing against land from formal financial institutions (rather than informal moneylenders). There was also a six-fold increase in fiscal revenue from land during the first 12 years of the project.[25] These factors contributed to the project's receipt of the World Bank Award for Excellence in 1997.

Land titling in Laos has played out against a dramatically different historical background. Very little land registration took place under the French colonial regime or during the war. Land was collectivized early in the socialist period, which began in 1975, but not to the same extent as in Vietnam or Cambodia. Land thus belongs to the state de jure but is held and managed de facto under an array of locally and culturally specific arrangements and customary practices.[26] The main push for land registration came after the 1986 policy shift to establish a market economy and the increase in investor interest in land-hungry projects such as plantations. Laos also has a much less established market economy than Thailand, and land markets are largely restricted to the main towns, their surrounding areas and rural areas accessible by road. Early titling experiments in Laos under the first phase of the programme (1997–2003) were limited to peri-urban areas around the capital city of Vientiane and selected secondary towns. Rural land titling started only in 2003, when the second phase of the programme was extended to rural areas on the Vientiane Plain and to some remoter multi-ethnic sites.

Land titling in Laos followed the model developed in Thailand, while seeking to take on board some of the critiques of the Thai project.[27] The critiques highlighted gender issues and the problem of dispossession (see below) but also made the broader charge that the Thai project was insufficiently attentive to the social implications of titling—or, indeed, to the fact that titling was a social and political process. For much of its early life, the Thai programme was implemented as a largely technical exercise, conceptualized in terms of training, technology transfer and administrative design. The problems that emerged were seen as technical in nature and included poor-quality maps, the holding of different kinds of title in different offices, and a shortage of surveying and land valuation expertise. When critiques of the Thai programme and the first phase of the Lao programme indicated that social aspects had been overlooked, LEI responded by funding a small independent study of the extension of titling to rural areas, including effects on customary land tenure, on inheritance and gender relations in different ethnic groups, and on land markets.[28] Over time, the managers claim, the Lao programme has become less rigidly prescriptive, more sensitive to existing social relations around land, and more accommodating of informal arrangements.[29]

Exclusion, as we noted, is intrinsic to titling, so the way titling is conducted has long-term implications for control over land. In Thailand there has been concern over influential persons colluding with corrupt land officers to gain control over land. The best-known case is in the district of Ban Hong, in the northern province of Lamphun, where 2,400 hectares were titled to outsiders without the knowledge of local farmers who had previously worked this land for many years.[30] Disgruntled farmers responded to the fencing off of their land by carrying out their own seizure, characterized as a "people's land reform", in which they occupied unused plots. More generally, title can give legal status to land that was grabbed in the past through dubious and sometimes violent means. This issue has not received much attention in Thailand or Laos, although it has been highly politicized in the Philippines, where titling is said to "'legitimize historical injustices' by favouring the rights of current landholders over those of the indigenous people who were moved from those lands in the past".[31]

Titling can intensify processes of land loss and accumulation by making land more transferable. Indeed, facilitating the operation of a land market is one of the goals of titling programmes, and proponents of titling measure their success in these terms, as we noted in the case of Thailand.[32] The point is not that titling transforms land into a commodity for the first time. Prior to titling, most agricultural land in Thailand was already potentially available for sale, or used as collateral for loans through informal, place-based networks. The difference introduced by titling is that land becomes saleable outside place-based networks, enabling anyone at all to buy and sell it. Titling's virtue, from the perspective of its promoters, is precisely its capacity to *delocalize* the land market and thereby eliminate the "stickiness" that might otherwise hamper transactions.

In both Thailand and Laos, the lubrication of land markets was declared to be a specific goal of the titling programme. However, the two projects approached the question of which land to title first in opposing ways. The priority in the early years of the Thai programme was to accelerate titling in disadvantaged, remoter provinces in order to stimulate their development. This decision strengthened the programme's claim to be a vehicle of national inclusion, a point picked up by de Soto when he visited Thailand in 2002 and applauded legal recognition of the assets of the poor as "a nation-building process".[33] However, it ran counter to the alternative argument, that land titling should follow the evolution of land markets and begin in areas where intensive land use had created the demand for fully institutionalized property rights.[34]

In Laos, there has been tension between the pressure for rapid coverage of the whole country (drawing on the World Bank funding) and a more considered approach based on trials in key areas (drawing on the AusAID-supported technical assistance). To date, the programme has followed the evolutionary approach and prioritized areas with existing land markets. As the programme was extended into rural areas, it gave priority to relatively accessible areas alongside roads that were already drawn into market processes, referred to in Lao as *khet sethakid* (economic zones). The independent evaluation team mentioned above confirmed that markets in land were indeed well advanced along rural roads by the time the titling programme arrived.

While the role of titling in facilitating land transactions is evaluated positively in neo-liberal terms, critics claim that making it easier both to sell land and to use it as collateral intensifies the vulnerability of poor people to dispossession by means of debt, a problem we discuss further in Chapter 6. The well-established finding that titling increases the market value of land is also evaluated differently by proponents and critics, with the former stressing the wealth effect for smallholders and the latter arguing that rising prices are on balance negative for farmers, especially younger generations who cannot afford to buy land at high prices. Ironically, far from increasing the efficiency of land use, higher values for titled land can lead to the speculative purchase of land and the suspension of productive activities (we say more about this in Chapter 5). In principle, tax incentives can help deal with this problem, and several countries in Southeast Asia have differential taxes on land used for agriculture versus alternative purposes. In practice, however, large areas of high-quality agricultural land that has been titled are held by absentees, and are unused or underused. This pattern was rare under customary tenure systems, where the protracted absence of the landholder, or failure to contribute to collective activities such as maintaining irrigation channels, could result in the land being quietly repossessed by kin and neighbours or formally reallocated.[35]

Debates about which land should be prioritized for titling purposes, and the connection that critics draw between titling and dispossession, carry implicit suggestions that some people and areas should be opened up fully to the market while others should be protected from it. Protective measures devised by people who do not share de Soto's faith that the magic of the market will lift everyone out of poverty have a long colonial pedigree in Southeast Asia and elsewhere.[36] But top-down and NGO schemes to protect people from markets are routinely accused of being patronizing, and they sit uneasily with discourses and values of empowerment, local wisdom and so on. Although critics of titling argue that it is the wealthy who are best able to take advantage of delocalized land markets while the poor are not interested in selling their land and hence have little to gain from land titling,[37] poor people often want the right to sell their land, even if they don't plan to exercise that right. More seriously, in a context where titling is becoming widespread, any land user without title is deeply insecure. The 2008 study carried out by Hirsch and others found that villagers were more upset that their common grazing lands, unlike the land immediately adjacent to their settlement, had not been included in the land titling, rather than being worried—as the research team had anticipated—that their pastures would be enclosed.

A case from southern Laos will serve to illustrate some of the direct and indirect ways in which the titling programme has excluded villagers from access to land.[38] Lak Sipkao is a village in Champassak province, strung out along the eastern side of the main north-south road between Pakse and the Cambodian border at the Khone Falls. This village is in one of the districts selected for rural land titling in Phase 2 of the Land Titling Project. Until 2007–8, lowland Lao and Alak farmers in this mixed ethnic village had been practising a combination of wet rice paddy cultivation on a very limited area of suitable land, and extensive rotational shifting cultivation to the east of the village.

In 2008 the Community Education Support unit arrived in Lak Sipkao to inform villagers about the survey and titling process. The Systematic Adjudication Team (SAT) arrived several months later and issued titles for the house plots only. No agricultural land was surveyed and titled. The reason for this omission was the incentive structure that rewards SATs based on the number of plots titled. Surveying residential land is much easier and quicker than surveying agricultural land, for a number of reasons. Plots are smaller, more accessible and well cleared. Making appointments with neighbours to certify boundaries is simpler. As a result, the SATs tend to move on once residential land, and in some cases wet rice land immediately adjacent to the settlement, has been titled.

Only around 70 per cent of the residential plot land was titled. A 25-metre-wide strip of land on each side of the highway was excised from the titled area, on the grounds that this would make it easier for the government to expand the road at a later date without having to pay compensation. Many villagers found that their houses were half inside, half outside the boundaries of their titled land. Others found that their land had been surveyed in such a way that they had no right of way to a road.

At around the same time as the residential plots were being titled, a Vietnamese rubber company was granted a long-term lease to the entire area on which villagers had been practising shifting cultivation, as well as to some of their already scarce wet rice land. The area was razed and a barbed wire fence built around it. Villagers were compensated for fruit trees planted on plots for which they had paid land tax, but they received no compensation for their fallow land. Some farmers had earlier planted teak in their swidden fields to enhance their security against state appropriation, but the strategy failed, as officials ruled that just a few teak trees per hectare did not qualify as productive land use. Officials also argued that the swidden land was public land, since it was not titled and most was uncultivated in any given year. So the lack of titles was combined with non-payment of taxes and inefficient land use to justify expropriating the swidden land and allocating it to the plantation, on which the villagers were expected to work as rubber tappers.

Summarizing these events, we see that the outcome of land titling in Lak Sipkao was to grant landholders exclusive use and disposal of their titled house plots. But the road easement is now state land for which villagers will not be compensated, and much of their swidden and some wet rice land has been expropriated for commercial estate agriculture. In the latter two cases, it was not titling but absence of title that legitimized the expropriations. Before the titling programme began, however, absence of title could not serve as a rationale for the state to expropriate land in the name of development. The lack of title added a layer to already existing frameworks (non-payment of taxes, inefficient land use, the state right to widen roads) that were already available to legitimate expropriation.

It is a premise of land titling that it does not create new rights, but rather recognizes and formalizes existing ones. The case of Lak Sipkao illustrates how both the presence of titles and the absence of titles do more than that, by applying uniform legal arrangements across a variegated space, while offering only selective recognition of previous arrangements (the house plots were recognized, the swiddens were not). Another arena in which new rights are inadvertently created and old rights undermined is within nuclear and extended families. There have been cases in northern Thailand, for example, where land that was managed in common by matrilineal kin groups was assigned individual title.[39] Often, there is a lack of correspondence between how customary kinship practices structure ownership and inheritance, and the assumptions about families embedded in national land and family law.

In Thailand and Laos, titling programmes have vested ownership in the head of household, a role legally defined as male. This has been problematic in Thai and Lao social contexts where land is passed from mother to daughter. Responding to critics, the Lao programme took gender issues on board and announced that it had granted more titles to women alone than to men alone, while the majority of registrations were in the name of both husband and wife. However, this approach has not necessarily brought titling in line with customary practice. Some women in patrilineal Hmong households have been included in ownership certificates in ways that run counter to Hmong tradition. When a Hmong man dies, it is his brother—not his widow—who inherits his property, so the presence of his wife's name on the title will either be ignored or it will create a point of tension and possible conflict.[40]

Although titling programmes are intended to clarify rights and eliminate ambiguities, these examples from Laos show how they can create a new level of confusion and opacity. The standardized rights represented by a title are often out of step with local practices that customary tenure regimes, in their infinite variety, readily accommodate. The gap has implications for local land relations, since old ways of doing things come under challenge but new methods are not fully established. It also has implications for legibility, as registers of title deeds may bear little relation to possession and the accompanying sense of legitimate entitlement that exist on the ground. Even complete titling, from this perspective, would not end history.

Land Formalization: Hybrid Practices

The priority placed on titling in contemporary Southeast Asia suggests that all land relations will eventually be governed by title as the ultimate bounded, unrestricted, individualized right to land. However, the persistence of diverse state-recognized rights to land tells us otherwise. Some of this diversity exists because states have not yet managed to survey and register all the land they would like to title. Formalizations short of title have often emerged from a recognition that limited bureaucratic capacity and the high cost of titling require states to settle for easier and cheaper processes. In such areas, full title is the ultimate goal. In other areas, however, state actors wish to give people in situ some kind of formal recognition that is short of title, deliberately setting limits on full transfer and use rights.

How can we explain this aversion to title in an era of neo-liberalism? The scene was set during the 1980s and 1990s, when it became clear to governments across the region that it was politically and logistically not feasible to evict millions of people who were living and farming in the state-claimed political forest. The compromise was to formalize these de facto smallholder land occupations so the state could benefit from increased revenue, legibility and control but retain the land's public status. Inter-agency battles over jurisdiction were part of the picture, but there were also principled arguments made by different actors—farmers, NGOs and experts of various kinds—to the effect that some land should not be fully integrated into the market. The examples of Thailand and the Philippines will help to clarify the context in which these arguments were made.

In Thailand up to 14 agencies were responsible for the allocation of certificates of different sorts on public land until the early 1990s.[41] The most generic public land certificates were issued by the Royal Forestry Department during the 1980s as utilization rights (the so-called *sitti-tham-kin*, or STK documents). These came with restrictions on use and on the area to be allocated to any single farmer. The purpose of the restrictions, in this case, was to encourage farmers to grow perennials on denuded forest reserve land and to create an egalitarian pattern of landholding in pre-emptive response to speculative land practices. At the same time, there was a rapid increase in the gazetting of national parks and wildlife reserves, demarcating land as strictly off limits to agriculture or extractive uses. During the 1990s, public land deemed appropriate for smallholder cultivation was progressively handed over to the Agricultural Land Reform Office and certificates were consolidated under a standardized ALRO certificate (known in Thai as *sor por kor*, or SPK title). Officially, SPK holders are protected from the market: SPK cannot be mortgaged or sold, although there is an active informal land market in SPK title. Ultimately the vision is to have all public land clearly demarcated between protected areas and agricultural plots, with the agricultural plots that were obtained through the land reform (SPK) retaining their special, protected status.[42]

A different combination of political dynamics and protective impulses is evident in two important land formalization projects in the Philippines: the Community-Based Forest Management Program of 1995 and the Certificates of Ancestral Domain Title (CADT) available to indigenous peoples under the Indigenous Peoples' Rights Act (IPRA) of 1997.[43] Since the Spanish colonial period, land and natural resources law in the Philippines has been based on the Regalian Doctrine, which states that all the country's land (with the exception of that covered by individual titles or state land grants) belongs to the state. Under this doctrine, the government has claimed roughly 60 per cent of the land in the Philippines as public domain. As one might expect, this land is not uninhabited, and by 1980 it was estimated that roughly 18 million people were living on it. Around this time, the government came to recognize that it "could ill afford (logistically, financially and politically) to evict even some of [these people] from where they settled, farmed, or collected forest products".[44]

The Community-Based Forest Management Program of the Department of Environment and Natural Resources (DENR) was one of several programmes established out of a desire to make "partners" out of upland people rather than treating them as illegal squatters to be evicted. Under the programme, upland groups can receive from the DENR a lease allowing them to occupy and use a certain area of forestland. While such a lease may look better than eviction, what stands out are the restrictions and responsibilities that accompany it. The land remains the property of the state; the lease is not in perpetuity, but rather only for 25 years (renewable), with the state reserving the right to cancel it for reasons of national interest at any time; the leaseholder is a "community", not an individual; and the lease carries with it a wide range of obligations with respect to the management of land and forests.

CADT available to indigenous peoples represent a genuine departure from the Regalian Doctrine. IPRA recognizes ancestral domain as the "private but communal property of indigenous peoples", thus not part of the public domain. IPRA has been hailed as a historic victory by advocates for indigenous rights in the Philippines and elsewhere, though others have pointed out the act's shortcomings. Again, what stands out are the striking differences between full title and CADT. To mention only three: ancestral domains "belong to all future generations", and no part of the CADT may be sold or otherwise alienated; the indigenous group holding the CADT has "the responsibility to preserve, restore and maintain the ecological balance within their territory and to restore denuded areas through reforestation and other development initiatives"; and the group is responsible for enforcing the boundaries of the domain.[45]

The references to community and communal property in the Philippine instruments for formalizing access to forestland reflect broader trends in land formalization in Southeast Asia. NGOs and other actors across the region have pressed for the recognition of communities, rather than individuals, as the proper recipients and managers of state forestland, even though this often sits uneasily with existing legal provisions.[46] This insistence on the communal has a number of provenances. Ideologically, many NGOs and, latterly, populist elements within state bureaucracies see communal tenure as inherently more egalitarian and less subject to accumulation than individual title. Politically, common property claims vis-à-vis the state or neighbouring villages may be more palatable than individualized claims over resources that are interpreted as selfish private interest.[47] Communal claims are also associated with the difficulty of parcelling up the management of forests and other common pool resources and hence the greater practicality of demarcation around "community" boundaries rather than individual household plot boundaries.

The difficulty with the argument for community forestland management, in much of the region, is that the "forest" land in question has long been farmed and occupied by smallholders. Indeed, the presence of farmers in these so-called forests was the original point of departure for formalization programmes, as we noted earlier. Hence, the "arborealization"[48] of their livelihoods is problematic. For many of the occupants of this land, it is farming, not forest management, that is the main agenda. On the tenure issue, some of these farmers are comfortable with communal models and, indeed, sometimes demand them, especially if they fit with existing customary land management practices. Other farmers, however, have long been accustomed to individualized tenure and fit awkwardly into communal tenure regimes. Nevertheless, these formalization programmes have given millions of people living in the political forest a level of security from eviction and a sense of inclusion in the national project that they did not enjoy when they were labelled squatters and thieves.

The striking finding that emerges from our examination of these formalization initiatives is the division of agricultural land between fully titled and restricted forms of tenure. The desired "end point" thus seems to be a system in which there are, in effect, two types of rural people. One group is composed of full market subjects, mobile, integrated into the national and global economies through land markets and other processes, and living in communities that have "succumbed", in Polanyian terms, to the social dislocations of capitalism. The other group are cushioned but also to some extent corralled. They could be said, in a de Sotovian vein, to be excluded from full participation in capitalism. The distinctions are to some extent ethnically based, with minorities and groups defined as "indigenous" edged towards the communal pole—but the distinction works only at a very broad level. More significantly, the classificatory schemes devised by land managers to distinguish between different types of people (e.g., indigenous/non-indigenous), or between types of land tenure (communal/individual) and land use (forest/farm), do not necessarily reflect the realities on the ground. An important implication of the gap between these schemes and de facto land relations is that non-titled land may not actually be protected from the logic of market processes and community dislocation. Rather, farmers find themselves in a hybrid situation in which they are producing for the market as individuals under the constraints of tenure systems that assume limited market engagement and prescribe collective management.

Land and Forest Allocation in Laos

National-level land formalization initiatives in Laos illustrate very well this emerging hybridity of land formalization. Full land titling in core areas, and in residential areas of some villages, coexists with conditional allocation in other, usually more peripheral sites. Together with the subsequent land titling programme, the Land and Forest Allocation (LFA) programme has been the main vehicle for putting this hybrid regime in place. The programme had its origins in the immediate post-socialist reform period of the early 1990s, when two decrees provided for the devolution of authority and responsibility for forest and forestland management to the local level.[49] The stated intentions of this devolution were forest protection and poverty alleviation through rationalized and more productive land use practices. The programme evolved, however, into a countrywide effort to formalize rural land occupation by fixing village boundaries and by zoning the space within village territories into categories such as conservation forest, protection forest, use forest, agricultural land, settlement area and so on. One goal of this formalization was improved legibility of village land use, an objective highlighted by the land zoning map boards displayed prominently in front of every village.[50] Within the agricultural and settlement zones, land is further demarcated into individual plots that are eligible for full title in villages where the land titling process has come in. In most of the country, however, the LFA lies outside the reach of the land titling programme, which, as we noted, focuses on peri-urban and more accessible rural areas.

The LFA has become an encompassing rural agenda as the range of actors involved has expanded beyond the foresters who initially drove the programme to include all departments with land and livelihood programmes at village level. When Decree 169 was issued, many observers saw it as a progressive move that would give local communities a recognized role in managing resources within their vicinity[51] and provide a local check on illegal logging.[52] However, as the project has proceeded, its negative implications have become increasingly apparent, even if there has been little debate or challenge to the programme in Laos.[53]

Three elements of the formalization programme implemented through the LFA are problematic, mainly because—like titles—they do not just clarify existing rights; they change relations of access and exclusion. First there is the geographical demarcation between titled land and land managed under the LFA. LFA was designed for remote forested areas. Land titling was designed for areas with established land markets. Increasingly, however, the two are coming into proximity and overlapping. Unlike in Thailand, where the boundary between land eligible for titling and ineligible political forest land has been demarcated and fixed,[54] many of the rural villages where the land titling programme is being implemented in Laos were previously demarcated and zoned under LFA. This was the case, for example, in the village of Lak Sipkao described above. There is no contradiction between the LFA and titling of individual house plots and nearby rice fields so long as the landholders remain within the community. Title, however, is an inherently "delocalized" licence-to-sell, as we have pointed out. So, once titled, plots could be sold to outsiders, excising the land (and the title holder) from village-based management that is the core feature of LFA zoning. The disarticulation between collective land management, on one side, and potentially individualized land ownership cemented through title on the other is one of the issues on which the Lao government and the land titling programme donors disagreed vehemently, ultimately leading to the programme's cessation in 2009.

The immediate disagreement was between the National Land Management Authority (NLMA), which took over the Land Titling Project from the Ministry of Finance in 2007, and the donors—the World Bank and AusAID—over procedural issues in titling. Specifically, the NLMA wished to link land titles with zoning and management plans, in effect reintroducing conditionality in land use as a prerequisite to issuing titles. This ran counter to the donors' neo-liberal notions of property rights in land as a separate issue from environmental and other regulations. There were also disagreements about the excising of land slated for development, such as the roadside land referred to in the Lak Sipkao case. At a more fundamental level, these issues reflected deep differences in the land titling programme's notion of land as private property versus the (only partly post-)socialist Lao government's view that all land remained state property (or what government often refers to in Laos as "national common property") and that titles were a kind of trusteeship. Interestingly, the land titling process continues in modified form in two pilot provinces under the NLMA preferred model with the support of the German aid programme GTZ, which has elsewhere supported linkage between land use planning and property rights issuance.[55] Meanwhile, titles already issued retain their legal standing, since they are recognized under the Land Law, but the original ambitious schedule of titling is not being followed. In effect, then, the hybridity of the land formalization system has been further entrenched.

Second, the LFA introduces rigid resource boundaries within villages, fixing practices onto spaces where previous land uses were often more flexible. It also creates more rigid boundaries between villages and sometimes results in one group being excluded from previously shared resources.[56] As with land titling, of course, the LFA is not completely effective in this regard. Keith Barney writes of his field site of "Ban Sivilay" that most villagers "do not pay much attention to the map, and few could articulate specifically how it regulated their agricultural or livelihood practices. Villagers do know, however, which forests are likely to result in a visit from the district or provincial forestry office if cleared for swidden fields."[57]

Swidden practices are at the heart of the LFA's third, and most critical, exclusion. The programme has become bound up with the state agenda of "stabilizing" shifting cultivation, which the government views as primitive and destructive. Eradication of shifting cultivation has previously involved attempts to resettle shifting cultivators from highland to lowland areas, efforts that often ran into problems of land availability and concerns that resettlement was forced. When highlanders did move, they were usually placed in new villages in close proximity to existing settlements, where the two groups had to negotiate over resource access and conflict was common. Hitching the LFA to stabilization helped solve these problems in two ways. First, stabilization meant that shifting cultivators could be left in situ, with the LFA system used to limit the area they could use for swidden cultivation. Usually, shifting cultivators are allocated a maximum of three plots and must shorten their fallow practices to a three-year rotation that is unsustainable under existing cultivation practices.[58] This limit obliges them to resettle "voluntarily" or to intensify their land use by adopting new crops, such as rubber. Second, for groups that "chose" to resettle, or those that had already been moved, the LFA was used to demarcate village territories as a way to clarify resource claims and thus pre-empt conflict between new and established residents. The assumptions behind this approach, however, are not always sound. In some cases, such as those where Hmong from the uplands have come to share natural forest with lowland Lao, formal demarcation has generated more conflict than previous informal arrangements.[59]

Despite these problems, the LFA has provided space for some interesting adaptations where village governance and leadership have been strong. One of these is an initiative to protect villagers from the risks presented by the land market, a protection designed and implemented, in this case, from below. Such an initiative occurred in a Khmu village in Luang Prabang, where the village committee prohibits the transfer of cultivable land to people from outside the village, in order to maintain the stock of land available for future generations to farm. Exceptions are made to this rule for teak smallholdings, where sale of the standing timber and the land under it is allowed, in recognition of its importance as a source of funds for villagers wanting to invest in big-ticket items such as tertiary education.[60] In a lowland Lao village in Champassak province, villagers decided to manage their rice farming land inside the LFA-demarcated area through a hybrid tenure arrangement.[61] During the wet season, farmers plant 164 hectares of rice in individual plots. During the dry season only 90 hectares of this is irrigable, and the village leadership stipulates that the land be re-apportioned so that any family wishing to plant is allocated land at a village-set rent paid in paddy to the owner. These flexible and need-driven land allocation arrangements rely on the LFA's clear demarcation of village land, but they would be undermined if the rice land was titled, in which case it would be alienable to outsiders.

These examples notwithstanding, the broader effects of the LFA have been starkly negative. An assessment of rural poverty carried out for the Asian Development Bank found that the single greatest contributor to a decline in living standards among the poor in rural Laos has been the LFA's restrictions on swidden rotations.[62] The effects have been particularly severe where villages have been involuntarily resettled, but even in situ "stabilization" has created widespread poverty, despite its billing as a scheme for poverty alleviation.[63] Moreover, the distinction between voluntary and involuntary resettlement is blurred when upland livelihoods are made unviable by LFA restrictions.[64] The

LFA thus illustrates the "slide" from formalization into resettlement. While other rural development programmes sometimes soften the blow, many people whose resource base has been curtailed now eke out an existence in lowland areas through begging.[65]

Land Reform and Land Settlement

Southeast Asia has a long history of land reform, with projects driven by different objectives. In some cases, for example in South Vietnam in the 1960s and the Philippines in the decades after World War II, redistributive land reform was pre-emptive, intended to defuse Communist insurgencies spurred by significant inequality in landholding. In other cases, Communist regimes instigated land reform programmes in the aftermath of revolution, notably in North Vietnam during the 1950s, when large landholdings were distributed to the rural landless and land poor. Yet elsewhere, land reform programmes have emerged from progressive social movements, when these were matched by an equally progressive government response. Efforts in the Philippines after 1986 are one well-known example; another is Thailand's Agricultural Land Reform Act of 1974, which established the Agricultural Land Reform Office in 1975 as a response to peasant and student mobilization.

States do not have a monopoly on land redistribution. Land occupations in several countries have seen the poor claim land on their own behalf. In Thailand, the "people's land reform" movement mentioned earlier has claimed land on behalf of the rural poor, with the best-known example in Lamphun province. Indonesia has an active land-reclaiming movement, described in Chapter 7. Usually the land reclaimed has been acquired by its legal owners through dubious means, and often it has been left idle. The Philippines stands out for its effective blend of mobilization for land reform from below, supported and legitimated by an official land reform programme.[66]

Land settlement is the second main redistributive programme in Southeast Asia, usually involving the grant of state-claimed land to settlers brought in from distant locations. Most countries in the region have had high-profile sponsored resettlement schemes. In Malaysia, the Federal Land Development Authority (FELDA) programme was a means to allocate resources to poor Malays and to establish estate agriculture over large areas of Peninsular Malaysia.[67] In Indonesia, the transmigration programme (discussed further in Chapter 7) has moved large numbers of poor farmers from the crowded islands of Java, Bali and Madura to the so-called outer islands of Sumatra, Kalimantan, Sulawesi and West Papua. Transmigration was a continuation of the Dutch "colonization" programme of the 1920s and shared its aims: to promote agricultural production, relieve population pressure and help consolidate the state's territorial control.[68] In Thailand, several government departments created sponsored settlement schemes, although their effect was dwarfed by the number of people who resettled spontaneously. In the Philippines, state encouragement of migration from Luzon and the Visayas to Mindanao continued through most of the 20th century.[69] Post-1975 Vietnamese programmes designed to move people from the crowded deltas to the Central Highlands will be discussed in Chapters 4 and 7.

Across the region, it has been much more feasible politically for states to carry out land settlement than land reform, and there is some blurring between the ways the two programmes are described. Except in the case of post-revolutionary North Vietnam in the 1950s, land reform has made few inroads into addressing inequalities in landholding in older settled areas in the rice-growing heartlands. The political obstacles to progressive land redistribution have been too great, and the loopholes in specific programmes too glaring, to achieve more than a superficial reallocation of land. In this context, the continued availability of land at the forest margin meant land allocation could be achieved by other means. Land settlement, sometimes described strategically as land reform on public land, has thus served as a safety valve that reduced pressure for more politically contentious land reform schemes. Thailand's Agricultural Land Reform Office (ALRO) has focused mainly on land settlement or on ex post facto recognition of spontaneous settlement of the forest fringe. When ALRO has attempted to allocate land, existing landholders have responded with violence and intimidation directed against the beneficiaries.[70] Similar problems have arisen in the Philippines, where the power of landlords and their willingness to protect their holdings with force have been major obstacles to land reform.

The Comprehensive Agrarian Reform Program in the Philippines

While calls for land reform and failed attempts at implementation were a constant refrain in Philippine politics during the 20th century, substantial reform began under the Marcos regime in 1972.[71] Marcos' programme represented a seemingly progressive policy response to rural inequality and unrest. The motivations behind the reforms were complex, however. A key goal (and one very much shared by the country's American advisers) was to head off the resurgence of leftist revolt, the underlying conditions for which had not disappeared since the decline of the Communist-inspired Huk rebellion of the 1950s.[72] Other influences were the experience of post-war land reform in Taiwan, South Korea and Japan, which development agencies increasingly saw as a model; the development orthodoxy of the Green Revolution, which claimed that smallholders farming their own land were more productive than large-scale farms; and Marcos' desire to undermine some of his own elite political opponents whose power base was in landed property. The success of the Marcos reforms was limited.[73] Only a small proportion of the targeted 1 million hectares was redistributed between 1972 and 1986 (the end of the Marcos regime), although about half the targeted area was converted from sharecropping to fixed rent leasehold status.[74] Tenancy reform rather than land redistribution was thus the main achievement at this stage.

The 1988 Comprehensive Agrarian Reform Program (CARP) introduced by the Aquino administration was much more far-reaching. Created in the heady period following the overthrow of Marcos, CARP differed from the earlier reforms in being driven by social movements protesting cronyism and the highly unequal and corrupt landholding situation in the Philippines.[75] The programme sought initially to redistribute roughly 10 million hectares of land, though the target was later reduced to around 8 million.[76] While the Marcos reforms covered a limited area of tenanted rice and corn lands, CARP was initially designed to redistribute more than 80 per cent of the Philippines' farmland. The means for redistribution included voluntary offers to sell (with a compensation premium), compulsory acquisition, and distribution of stocks in land-based enterprises.[77] Government subsidies helped cover the gap between what poor farmers could afford and compensation for those relinquishing their land.

Formidable powers have been arrayed against land reform in the Philippines, notably the political dominance of landlords from elite families and the loopholes that allowed spurious division of landholdings among relatives, cronies and clients. CARP implementation on the estates of entrenched landlords/warlords has hardly begun,[78] despite some high-profile successes such as the redistribution of lands owned by the Aquino family in Quezon.[79] Nevertheless, by 2004 the government claimed that it had distributed 5.8 million hectares—58 per cent of the nation's farmland—to 2.7 million CARP beneficiaries, or 44 per cent of the total rural population.[80]

Although the numbers just quoted are impressive, on further inspection CARP has had some important limitations, and many people who should have been included as beneficiaries have in fact been left out. First are the tenant farmers and landless rural poor who have not been able to access the programme. Counter-intuitively, this group includes some people who could have presented themselves as beneficiaries but did not. Their reasons are complex but often have to do with a perceived loss of security, as land reform parcels land, fixes boundaries and creates own-account landholders, each of whom must fend for themselves. Communities of estate workers and tenants are broken up by land reform, and with them the myriad kinship-based and other social relations that bound farmers and workers to the land but also to each other, albeit under highly unequal social conditions. With violence and poverty pervading the countryside, many plantation workers and tenant farmers have opted to retain the protection of their landlord/patrons and their low-wage but relatively stable jobs.[81]

The second group excluded are the people officially counted as beneficiaries who have not been able to take possession of their land, or had the titles they were awarded reversed due to legal challenges, or gained only nominal possession because of lease-back or "joint venture" arrangements that left the original landlords in control. A third group comprises the people from whom the land was seized initially; their claims are not recognized by CARP, which is not designed for this kind of historical redress. On the contrary, as Borras points out, the "willing seller" model of land reform legitimizes the position of landlords, however fraudulent or violent the process through which they came to hold the land. Absentee landlords are entitled to the same compensation as owner-operators.[82] This lack of a social and historical perspective—a concern with who the people are, and how they came to be in possession of land—is also a characteristic of land titling programmes, as we have seen.

Fourth, and more contentiously, it could be argued that land reform beneficiaries are excluded from a significant right held by many other smallholders in the Philippines: the right to participate in the land market. This exclusion is not unique, since it is applied also to people covered under indigenous land claims and community forest programmes, as we earlier explained. To understand why this limitation figures in CARP, it is useful to compare the policy rationales for CARP and for land titling. Both programmes are promoted as vehicles to increase farmer efficiency through the granting of clear and exclusive rights to a plot of land. Like titling, CARP was meant to make micro-entrepreneurs of the Philippines' millions of poorer farmers. The key difference, however, is that while titling programmes see increased land sales in positive terms, CARP seeks to hold capitalism at bay. CARP beneficiaries are to be entrepreneurial, but they will not accumulate land, nor will they sell it.

CARP's tough rules forbidding land sale emerged from the experience of the Marcos-era land reforms, when there were rules forbidding sale but studies showed that they were routinely violated or circumvented.[83] As in the land formalization programmes discussed above, then, CARP tries to protect land reform beneficiaries from losing their land through debt, mortgage and sale, the "everyday" mechanisms of dispossession we discuss further in Chapter

6. Among proponents and critics of CARP, there is continuing debate over whether the effort to protect beneficiaries from the land market is beneficial or effective. Some argue, for example, that every farmer needs credit to develop their enterprise, so CARP land should be eligible for use as collateral for loans. Others highlight the attendant risks.[84]

A final caveat to the impressive number of land reform beneficiaries cited earlier is the extent to which the land distributed has been public rather than private. The Department of Environment and Natural Resources (DENR) administers the CARP programme on public lands in two main components: Alienable and Disposable Land, and Community Based Forest Management. By 2004, 72 per cent of the land transferred under CARP was public, and much of it, moreover, was already cultivated by smallholders or secured by larger landed interests. These numbers indicate a significant slide from land reform to land formalization—using the land reform programme to regularize the tenure status of the people who are already in place. Some critics argue that any land reform that does not remove land from landlords is not genuine reform or progressive, but defenders of CARP argue otherwise. They point out that much public land is already de facto privatized, as it is held by agribusinesses or large farmers who have acquired leases at nominal cost or through extra-legal or ambiguous legal arrangements. So redistributive land reform is needed as much on public land as on private land, and there is plenty of scope, since the land areas involved are significant.[85] Sadly, however, wresting away this land for distribution to tenant cultivators and landless labourers has often proven to be just as difficult to carry out as reform on "private" land.[86] DENR on behalf of CARP did not transfer individual access to all the smallholder-farmed public land. Instead, it allocated differential land rights according to the land classification. It allocated individual land rights to 1.3 million hectares of public land classified as "alienable and disposable", which in the Philippines means that they are no longer considered part of the forest domain. However, as objects of the CARP programme these allocated lands are under individual possession but cannot be sold. DENR allocated a further 1 million hectares in the name of CARP under CBFM 25-year contracts available through the community-based forest programme discussed above.[87]

Conclusion

This chapter has described a remarkable set of state-led interventions that have transformed land relations across the region, beginning in the 1950s and intensifying since around 1990, when the range of interventions expanded and old ones were revived. Our chapter title, "Licensed Exclusions", highlights the role of the state and the use of regulatory power as the main mechanism to define who shall hold what land, under what conditions. Regulatory power is inherent in the formalization of claims on land and embedded in the inscription of claims as legally enforceable, state-recognized rights. Regulation is also the power behind the land codes and decrees that establish conditional rights to dispose of land, to manage it under particular zoning classifications, and so on. These regulatory exclusions operate at different scales. Torrens title bestows a state recognition of ownership that supersedes any customary claim to a piece of land and gives the individual title holder a perfected right to exclude other users and other claimants. State regulatory power has also been used to formalize landholdings at the village level in Laos, where the Land and Forest Allocation programme attempted to create a boundary between forest and farm, and between one village and the next. State regulatory power also delimits which areas of national territory are to be open for full titling, which areas are off limits to human settlement and agriculture, and which in-between areas are subject to conditional rights of occupation and use. As state regulatory powers over land relations have expanded and intensified, customary powers of regulation have been diminished, although they remain locally effective in some areas, where they sometimes undermine state regulatory systems (making titles opaque, for example) and sometimes operate within them (as when Lao villagers adapted the LFA creatively to meet their own goals).

While licensed exclusions are put into place primarily through regulation, other powers are simultaneously in play. The market is one of them, since cost remains an obstacle to the wider issuing of titles. At an individual level, titling is an expensive affair. At the level of subsidized national programmes, state efforts to provide title have been limited by the huge cost of surveying and registering every plot of a country's agricultural land. Even when titles have not been issued, as in much of Laos, they have still caused changes in land relations, in the sense that the existence of title in some places delegitimizes occupation and use of land in others.

Proponents of title see land as an asset that should be in the hands of (some of) the rural poor, who will thereby become full market citizens, equal players in the nation's economy. The power of the market is thus substantive in providing the mechanisms (mortgage and foreclosure) through which land alienation occurs, and it also serves as a legitimating device, providing the rationale that attracts development funds and government commitments to land titling. Complete coverage of the national land mass under title would make the regime of exclusion perfect, and perfectly inclusive: in principle, everyone would have the right to exclude. In practice, however, many people would find themselves excluded from access to land. Recognition of this fact has led regimes across the region to devise a range of programmes to redistribute land where access has become unequal, or to prevent people from losing their land in the first place by limiting the operation of land markets. Thus, the idea of a "free" market in land, as the goal to be accomplished or the nightmare to be avoided, is a power operating more or less explicitly to frame all the programmes we have discussed.

The power of legitimation is equally ubiquitous in licensed exclusion, because in a matter as important as access to land, every form of exclusion and access requires a rationale. More important, deliberate state-driven interventions to change land relations need to be justified by both techno-scientific arguments (feasibility, efficiency) and moral ones (what is right). Legitimations are subject to discursive shifts. Early land reforms were justified in land-to-the-tiller terms as a solution to the problem of poverty: people were poor because they lacked access to land. This idea morphed into a discussion about the need to make land access secure through formalization or titling, in one iteration, or by pinning villages or communities into place, in another. Migrants to the forest edge, or shifting cultivators, so the argument went, would be less poor if they were settled through programmes of allocation, zoning and so on. As land reforms came up against intractable opposition, land settlement programmes emerged that repeated one of the main pillars legitimating land reform (poor people need land) but targeted public land, which was easier to distribute than land held by landlords. At the end of the 20th century, land titling came to replace land reform as the main vehicle for poverty alleviation, at least among the experts generating rural development programmes. Yet despite the transformations in legitimating discourses we have summarized here, a remarkable feature of Southeast Asia is that all four approaches—titling, formalization, settlement and reform—are still being actively implemented in the region.

The evidence for the working of our fourth power, violence, is mixed. Titling provides legal rights to land and holds out the promise of replacing exclusions backed by fear and patronage with a fair, rule-governed and peaceful system of adjudicating the right to exclude. Yet, the very prospect of achieving title has also provoked extrajudicial land-grabbing in anticipation of state recognition. Violence has also skewed the implementation of programmes such as CARP in the Philippines and ALRO in Thailand, where many farmers have been unable to occupy land that is legally theirs due to threats by thugs. The juxtaposition, and sometimes the mingling of licensed exclusions and force, will come sharply to the fore in our discussion of the post-agricultural land developments in Chapter 5. Next, however, we turn to the transformations in land relations brought about by ambient environmentalism.

Notes

1. This point is vividly brought home by maps that depict the cultivated areas of what is now Thailand in the 1850s, 1950s, 1980s and 1990s, in Phongpaichit and Baker 1995: 5–6. See also De Koninck 2003: 218; Scott 2009.
2. Trade policy changes include the free-trading Bowring Treaty of 1855 between Britain and Thailand, and the near-simultaneous opening of key ports in the Philippines to international trade.
3. Scott 1998.
4. For an example from the Philippines, see Wenk forthcoming: 15.
5. De Koninck 1996.
6. On Laos, see Barney 2008; on the Philippines, Wenk forthcoming: 84–6.
7. Hirsch 1990a, b.
8. See Wenk forthcoming: 19–23.
9. See Sjaastad and Cousins 2008: 2.
10. Fukuyama 1992.
11. On this dynamic in North Borneo, see Doolittle 2005: 33–4.
12. Kratoska 1985, Wolters 1999.
13. Rattanabirabongse *et al.* 1998.
14. Hutchison 2008: 335.
15. See McAuslan 2003.
16. de Soto 2000.
17. See Hall 2004a: 404.
18. Hughes 2004.
19. For Macapagal-Arroyo, see Hutchison 2008: 342 fn 12; for Yudhoyono, see Fauzi 2009; for Thaksin, see *Nation*, 8 Nov. 2002.
20. Deininger 2003.
21. The empirical material for this section is drawn largely from the "grey literature" of the land titling projects, NGO critiques and two small research projects carried out during 2003–4 and 2008 in Laos in which Philip Hirsch participated. Unless otherwise indicated, material on Thailand is based on Rattanabirabongse *et al.* 1998, and material on Laos is based on Fujita *et al.* 2006 and on unpublished results from the study on which their publication is based.
22. Hirsch 1990a. See especially Chapter 2.
23. Rattanabirabongse *et al.* 1998.
24. Ibid.
25. Ibid.
26. On historical and current land tenure issues in Laos, see Ducourtieux, Laffort and Sacklokham 2005.
27. For a critique of the Thai programme, see Leonard and Narintrakul na Ayutthaya 2003. See also Rusanen 2005, Lohmann 2002.
28. Philip Hirsch was involved with the design and implementation of this study during 2003–4, the findings of which are summarized in Fujita *et al.* 2006 and in a further study of the Lao land titling programme during 2008.
29. For a discussion of new directions in land titling, see Dalrymple, Wallace and Williamson 2004.

30. Leonard and Narintrakul na Ayutthaya 2003.

31. Hutchison 2008: 336.

32. For example, see Rattanabirabongse *et al.* 1998.

33. *Nation*, 8 Nov. 2002.

34. Hayami, Quisumbing and Adriano 1990; Deininger 2003.

35. See Thangphet 1993.

36. Colonial and contemporary attempts to protect smallholders from land markets are discussed in Li 2010b.

37. Leonard and Narintrakul na Ayutthaya 2003.

38. This case is based on fieldwork carried out by Philip Hirsch during 2008.

39. Ganjanapan 2000.

40. Fujita *et al.* 2006.

41. Hirsch 1990a, b.

42. Hirsch 2009; Pongpaichit and Baker 1998: 255.

43. This paragraph is based on Wenk forthcoming: 49–56, 62–5. See also Chapter 7.

44. Wenk forthcoming.

45. Wenk forthcoming: 63–5; Li 2010a.

46. Ganjanapan 2000.

47. Tubtim and Hirsch 2005.

48. Walker 2004.

49. Decrees 169 1993 and 186 1994 cited in Baird and Shoemaker 2007; see also Phanvilay, Tubtim and Hirsch 1996.

50. Ducourtieux, Laffort and Sacklokham 2005. See also Barney 2008: 98–9.

51. At the time, the recognition of community management rights in Lao forestland policy was compared favourably to the Community Forest Bill under discussion in Thailand. See Vandergeest 2003a.

52. Baird and Shoemaker 2007.

53. Ducourtieux, Laffort and Sacklokham 2005; Vandergeest 2003a.

54. One problem facing the land titling project in Thailand has been the risk of titled and forest reserve land overlapping, and titling has thus been avoided in unclearly demarcated boundary areas. See Burns 2004.

55. For a further description, see http://hrdme.wordpress.com/2009/06/08/germany-supports-land-project/.

56. Tubtim and Hirsch 2005.

57. Barney 2008: 99.

58. Ducourtieux, Laffort and Sacklokham 2005.

59. Hirsch, Phanvilay and Tubtim 1999 describe such a case in upper Vientiane province. For a case from the Philippines, see Wenk forthcoming: 84.

60. Rob Cramb, pers. comm.

61. Findings derived from fieldwork by Philip Hirsch.

62. Baird and Shoemaker 2007.

63. Ducourtieux, Laffort and Sacklokham 2005.

64. Baird and Shoemaker 2007.

65. Vandergeest 2003a.

66. Feranil 2005: 279. See also Kerkvliet 1993; Peluso, Afiff and Rachman 2008: 388.

67. Sutton 2001.
68. See Elmhirst 1999.
69. Wenk forthcoming: 17–26; Abinales 2000.
70. Hirsch 1990a.
71. Abinales and Amoroso 2005.
72. Kerkvliet 1977.
73. Borras 2001: 550.
74. Borras 2001.
75. In 1998 the Gini coefficient of land was 0.61. Borras 2001: 548.
76. Borras 2001: 551.
77. Borras 2001: 552.
78. Bello *et al.* 2004.
79. Borras 2006b: 133–6.
80. See contrasting figures on land access and CARP generated by the World Bank, cited in Bello *et al.* 2004: 33.
81. See Hirtz 1998: 257–8. See also the ethnographic account of conflict among contending groups of beneficiaries in Rutten 2010.
82. Borras 2006b: 137–8.
83. Hirtz 1998: 257.
84. Bello *et al.* 2004.
85. Borras 2006b.
86. Borras 2006a.
87. Borras 2006b.

3

Ambient Exclusions:
Environmentalism and Conservation

In Southeast Asia during the 1990s and 2000s, discourses promoting conservation have become ambient—as pervasive, and as unmarked, as the air we breathe. Not only is a conservation rationale inserted into tenurial arrangements and controls on how land is used; it shows up routinely in national development plans and programmes for communities, and it spills out from within big projects such as dams whose environmental impact is a key concern. Our goal in this chapter is to explore how environmental rationales and conservation ideology—priorities that are usually framed in terms of inclusive notions of the common good—promote exclusionary outcomes not only in the more obvious sites such as conservation zones and protected areas, but in other arenas too. Legitimacy and regulation are the main powers at work in these exclusionary regimes, but force also plays a role, as does the market, as systems of incentives are devised to shape policies and practices at multiple scales.

The exclusions that concern us in this chapter are motivated by efforts to attain the "common good" promoted by conservation projects and discourse, with "common" defined at community, national and/or global levels. The main "good" is the indefinite preservation of life-sustaining natural processes. But the good comes at a price, notably the exclusion of smallholders from large areas of potentially arable land, and other limitations on land and resource use practices. This exclusion presents a conundrum most obviously for the conservation agencies associated with "integrated conservation and development programs", whose upbeat rhetoric is belied by a litany of project failures that tells of incompatibility between territorialized conservation and smallholder interests. It poses a dilemma for national policymakers and international development protagonists who draw on poverty alleviation and sustainability discourses in the same breath. It also causes divisions, if not actual dilemmas, among smallholders, some of whom clearly reject the constraints on their farming practices imposed by conservation agencies, while others participate in conservation-inspired projects for a range of political, economic and other reasons that are discussed further below. More generally, exclusions legitimated by conservation bring to the fore the competing value systems of ethnic minorities living in upland areas, urban middle classes, and rural smallholders, each with their own views of how forests, rivers and other elements of the natural environment should be used and allocated.

A primary power behind the implementation of conservation-based exclusion is the zoning of certain areas of land as off limits to agriculture through the drawing of forest boundaries, together with the formalizing of land settlement, processes we discussed in the previous chapter. Table 2.2 (in Chapter 2) shows the area of land in different Southeast Asian states that has been gazetted as forest domain, and within this the proportion classified as conservation or protection forest. There are often further subdivisions, designating the areas intended for "production" (logging) or "licensed" conversion to agricultural use through formally approved plantation or settlement schemes.[1] Most states in the region endeavour to restrict or deny smallholder access to any part of what Nancy Peluso and Peter Vandergeest have called the "political forest", that is, "lands states declare as forests".[2] In practice, the gaps and contradictions that exist between administrative categories, the actual priorities of state actors, and local uses and understandings of forest territory have been the source of a great deal of conflict, often dating back to the colonial period when forest boundaries were first declared.[3] Although a significant part of the state-claimed political forest is, in fact, farmed, the official "forest" designation imposes a condition of more or less chronic insecurity on the tens of millions of people who live and/or farm in the "forest" zones.[4] People living on land designated for the highest level of protection—conservation areas and national parks—are especially vulnerable to eviction and acute loss of livelihood. Their jeopardy has increased markedly since the 1980s, when donors started to provide the financial, technical and other resources needed to enforce park boundaries, transforming the often-nominal exclusion of the political forest into real exclusion on the ground. We show the forest areas set aside for protection and conservation in Figure 3.1.

While tensions between parks and smallholders are well documented, other intersections between land tenure and conservation may be less well known. Even within parts of national territories that are recognized as legitimate agricultural domains, zoning of land use at various scales imposes numerous constraints. For example, watershed classification programmes stipulate allowable patterns of settlement and land use based entirely on

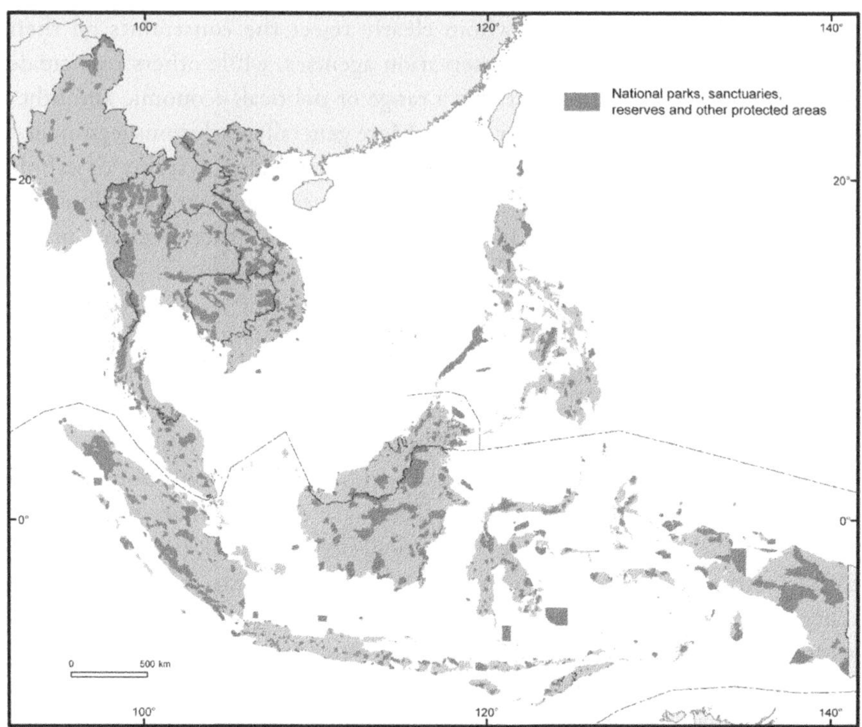

Figure 3.1 Southeast Asia: Protected areas (existing and proposed), 2009

Sources: Digital Chart of the World, World Database on Protected Areas.

physical criteria such as topography and past levels of vegetation disturbance. Some landholders face species-specific restrictions on harvesting wild products; others are restricted in their cultivation and extraction practices to fulfilling their "subsistence" needs. One of the most widespread and long-standing means by which states assert control over population and territory across Southeast Asia is the attempt to "stabilize" the uplands by prohibiting swidden agriculture and other customary land and forest uses, corralling settlement, and, in the more extreme cases, forcing people to resettle in the lowlands. Environmental concerns are the "ambient" legitimation for all these exclusions, even though other motivations—national security, or grabs for timber or land for plantation development—also play a role.

The study of conservation-inspired exclusions and restrictions has brought a new dimension to research on agrarian change that was largely absent before 1980, when studies were preoccupied with class differentiation and its attendant politics. The shift in research interests is associated with a geographical shift in the frontier of resource struggle. While core agricultural areas were the paradigmatic site of conflict over land in studies of agrarian transformations in Southeast Asia during the 1960s and 1970s, since then the focus of research on land conflict has moved to the uplands and other peripheral areas (such as coastal mangrove forests), where popular claims for land and livelihood run up against state-backed efforts to close the forest frontier. Arguably, these new struggles also have a class dimension, but the focus of the struggle has changed. While during the period 1930–75 Communist movements were actively engaged in mobilizing peasants against a landlord class, since the 1980s farmers on the forest frontiers, together with their allies among scholars, progressive social critics, NGOs and social movements, frequently encounter the state as the principal adversary, sometimes supported by transnational bodies concerned with the fate of the forest.

In addition to their vertical dimension—people versus state—environmental concerns also bring neighbouring villages into confrontation with one another, either through competition for land-based resources or sometimes through conflicts between lowland and upland communities over water and perceived erosion impacts. Since the national and transnational forest governance regime draws heavily on the discourse of conservation to legitimate exclusion, farmers seeking to gain or maintain access to land for agriculture must engage with this regime on these terms. They may counter with an argument that conservation agendas should not trump immediate livelihood needs; or they may argue that would-be farmers are also competent environmental stewards, fit to be included in forest management regimes, points we take up in Chapter 7. Either way, ambient environmentalism configures land struggles in crucial ways that will only intensify as climate change moves higher up the global agenda.

In the discussion that follows, we explore three different arenas in which conservation-driven, territorial exclusions redefine access to land for particular groups defined by their social characteristics or spatial locations. The first arena involves perhaps the most straightforward conservation-based exclusion, in that it is expressly about keeping people away from nature. The exclusionary effects of protected areas have become an important subject of struggle, and they have helped to split the conservation movement in many countries between "green"-inspired conservationists and livelihood-oriented groups that reject heavy-handed protected area demarcation, especially when this entails forest bureaucracies marking out territory to entrench their own jurisdictions. We examine these dynamics in the case of an Indonesian national park where exclusions are bound up in the interests of state actors, international donors, and a diverse array of local resource users and claimants.

The second arena, community-based natural resource management (CBNRM), is more counter-intuitive, since it involves a degree of self-exclusion. The founding assumption of CBNRM is that the people who live in close proximity to a resource should be directly involved in its management. CBNRM is deliberately inclusive, as it emphasizes common property, community cohesion and locally driven, collective action. Yet regardless of who makes the rules, natural resource "management" often comes down to restricted access: boundaries must be drawn around territories, and around social groups, separating those who may use a resource from those who may not. We draw our example from Cambodia, where CBNRM initiatives have been especially prominent and even livelihood-oriented conservation has had exclusionary effects.

Finally, we examine the insertion of green agendas into corporate projects, which are increasingly justified in terms of the "triple bottom line": they are meant to achieve not only financial but also social and environmental goals. Our case study involves a large hydroelectric project in Laos, proponents of which countered the challenge that it would be environmentally destructive by arguing that it would be good for conservation. Slipped into this nimble turnaround, and barely attracting notice, were the villagers who would have to be excluded from forestland. As corporations are drawn into the ambient environmentalism Michael Goldman dubs "eco-governmentality", they participate alongside states, transnational NGOs, development banks and other actors to regulate people's conduct through zoning, certification, resettlement and other means that are sometimes direct, but often unannounced.[5]

Protecting Parks from People

The most clear-cut example of conservation-inspired exclusion impinging on land tenure and land use is the establishment of protected areas that set large swathes of land off limits to most types of human activity. Protected areas include national parks, wildlife sanctuaries and other territories zoned for biodiversity conservation. In 1950 there were about 1,000 national parks and other protected areas worldwide. Half a century later there were between 29,000[6] and more than 100,000,[7] depending on which of the World Conservation Union's five categories of protection is taken as the criterion. These conservation zones cover more than 10 per cent of Earth's land area, and between 10 per cent and 25 per cent of the territory of most countries in Southeast Asia (see Table 2.2). Given the vast areas of territory worldwide that have been deemed "biodiversity hotspots" but are not yet protected, the process of territorially demarcated exclusion for conservation is potentially far from complete.[8] In the global South, protected areas are often (quite accurately) seen as beleaguered and as the victims of encroachment by development interests that place multiple demands on forested lands for agriculture, hydropower, road access, and other activities generated by demographic and economic growth. As we show in this section, however, they are also enormously important generators of exclusion.

The history of protected areas in Southeast Asia begins in the colonial period. Indeed, protection and conservation of forest has gone hand in hand with colonial and industrial forestry, and both are associated with the establishment of state control over forests.[9] Peter Boomgaard suggests that in colonial Java there was a symbiosis between conservation and exploitation, in which conservation maintained the resource while revenues from timber financed forest protection.[10] Even in uncolonized Siam, the British influence on forestry was associated with protection as well as exploitation of teak. The creation of national parks and other protected areas designed to preserve nature in perpetuity, rather than for future exploitation, is a post-war phenomenon in the region. Thailand's Wildlife Conservation Act (1960) and National Parks Act (1962) aimed to protect areas of high wildlife and botanical value by shutting out human activity (including the activity of people already living in the area) and led to the creation of the region's first national park—at Khao Yai, in 1962. Protected areas have expanded rapidly in the region since then, and their rate of establishment has accelerated since the 1980s. In the lower Mekong countries of Thailand, Laos, Cambodia and Vietnam, the figure is as high as 22 per cent—the largest proportion of national land area demarcated for conservation zones of any region of the world.[11]

The expansion of the region's protected areas has been driven by domestic and international influences and actors. To some degree, park creation derives from a global norm that says having national parks is now part of the "identity" of a modern state; for a country not to have any would be peculiar. These global influences have been intensified in many countries by the Western education of key officials. In Thailand, the US training of many of the country's senior foresters helped infuse protected area management with the wilderness ideal. The Thai environmental movement received an early boost from a former hunter turned ornithologist and conservationist, Dr Boonsong Lekagul, who founded the Association for the Conservation of Wildlife as early as 1950 and successfully campaigned for the early 1960s legislation noted above. These calls dovetailed with the interests of certain bureaucratic actors, particularly the forestry departments, which, across the region, have taken on the mantle of territorialized conservation. Domestic calls for nature preservation have intensified as urban middle classes with positive views of biodiversity conservation and nature tourism have emerged throughout much of the region. Their desires receive additional support from the presence in a number of countries of parks that have become icons of national identity. On the financial side, development aid and the participation of international institutions and NGOs have played a vital role in park establishment in some countries. For example, the Tropical Forestry Action Plan exercise during the late 1980s and early 1990s identified areas of forest that were to be protected against people through the creation of protected areas in Laos, Cambodia, Vietnam and Indonesia.

Protected areas exclude most directly when their establishment forces people to move. In an article subtitled "Evictions in Eden", Charles Geisler describes people who have been forcibly displaced from their homes and land by the reservation of territory for nature conservation as "conservation refugees".[12] Estimates of the numbers affected by such exclusions vary wildly, but the literature on Southeast Asia is full of examples. Park boundary demarcation has often either ignored existing settlements, or has excised settlement lands in such a manner that the villagers' basis for deriving a livelihood from surrounding land- and forest-based resources has been undermined. Thailand's Khao Yai National Park is one such example.[13] In Cambodia, similarly, World Bank and Global Environment Facility support for the Virachey National Park, in the northeastern province of Ratanakiri, has excluded indigenous minority people from access to land and forest resources.[14] Geisler makes the important point that while protected areas are often put forward as "an antidote to development and a bulwark against its externalities", there are in fact uncomfortable analogues between the displacement impacts of conservation areas and development projects.[15] He also claims that research and advocacy regarding "conservation refugees" has lagged behind that on people who have been displaced by dams, mines and large-scale infrastructure.

It is not only by expelling people, however, that parks exclude. Geisler argues that most of the people who were living in protected areas prior to their demarcation are still in situ, often in precarious legal situations and under more or less permanent risk of eviction. As many as 136 million people worldwide may be in this position. In Southeast Asia, too, few of the people living illegally in political forest have actually been evicted. Walker and Farrelly estimate that fewer than 0.5 per cent of the 1 million people living in northern Thailand's forest reserve areas have been forcibly removed from their villages,[16] although responses to this analysis suggest that this is an underestimate.[17] This is so in part because many protected areas are "paper parks", in the sense that geographical demarcation was not matched by budgetary allocation or management programmes on the ground. Like the state efforts to formalize and allocate land that we analyzed in Chapter 2, state conservation projects that have enormously ambitious goals on paper often end up having much less impact in reality. Again as in Chapter 2, however, the consequence of these failures is usually not to leave people's lives untouched by these projects, but rather to place them into an opaque and precarious situation in which different principles of resource regulation overlap, state priorities may be enforced unpredictably, and state (and other) actors have ample opportunities to extract bribes or worse.

The assumptions about the relationship between parks and people that have underpinned dispossessory approaches to conservation have, of course, been challenged, and alternatives have been proposed. In the international parks and conservation literature and community of practitioners, there is a fierce debate over the rights of nature versus the rights of local communities. Some authors have called for a new approach based on "ecologies of coexistence", a concept that challenges the very basis for excluding people from conservation areas.[18] A similar point is made in an unusually critical review of conservation and displacement in a book published by the Wildlife Conservation Society, an organization whose key staff members have had little sympathy with the idea of spatial coexistence between people and nature.[19] In the volume's introductory chapter, a prominent conservationist and an academic advocate of devolved rights and control over natural resources jointly raise questions about the efficacy of exclusion on both conservation and social grounds. They argue that unless the inequities of conservation-induced displacement are addressed, there is little future for conservation by this means.[20] These perspectives are somewhat unusual, given the fissure that has generated vitriolic exchanges between conservationists who see people as a threat, and for whom any concession to livelihood is warranted primarily in terms of its conservation-promoting diversionary effects, and those for whom the livelihoods of the poor are the main raison d'être of programmes that link conservation and poverty alleviation.[21] The debate over park-based conservation versus local livelihoods is endemic throughout Southeast Asia and beyond, and we illustrate here with just one case among scores of potential examples with similar dynamics.

Lore Lindu National Park[22]

The history of Sulawesi's Lore Lindu National Park illustrates the complex array of actors and agendas involved in protected area management. The park was announced in 1982, at the height of the Suharto regime, as a high-profile international gesture at the World Parks Congress in Bali. After a decade in bureaucratic limbo, it was gazetted in 1993, when the park's boundaries and its 229,000-hectare extent were confirmed and the Forest Department began to make arrangements for implementation. The New Order regime was earning its developmental credentials by this time, and the gazetting of Lore Lindu offered the chance to reinforce the regime's modernizing reputation by balancing rapidly accelerating economic growth with a demonstration of concern for the environment just a year after the United Nations Conference on Environment and Development (the Rio Summit). The park was framed within the global discourse of biodiversity conservation as, simultaneously, a "protected area", "Man in the Biosphere Reserve", "Center of Plant Diversity", "Endemic Bird Area" and "Global 2000 Eco-region". The main threat to all of this diversity, as early planning studies confirmed, was the local population involved in farming and rattan collection inside the park borders, as well as the inflow of migrant farmers from other parts of Sulawesi into the area. The latter problem became more acute as roads improved access and a lucrative new cash crop, cacao, made this hitherto rather remote highland area very attractive. The park was represented by its proponents as a last chance to avoid destructive human occupation by keeping people out.

Much of Sulawesi's Central Highlands had, in fact, been declared off limits to agriculture long before the park's creation. Under the Forest Law of 1967, the state had asserted its claim to the area under the jurisdiction of the Ministry of Forestry. In practice, however, tens of thousands of farmers continued to use this political forest land until their presence interfered with other plans: logging, plantations, hydro development schemes, resettlement programmes and national parks. Before 1993, when park authorities began to demarcate and enforce boundaries, people continued to engage in swidden agriculture, expand coffee stands, and collect forest products in the area that would become the park.

The conservationist discourse behind the park turned farmers into encroachers and collectors of forest products into delinquents, and legitimated the exclusion of people from an area of land that comprised between 13 per cent and 56 per cent of the affected sub-districts. This was the first and simplest exclusion that the park brought about. The challenge faced by state gazetteers and managers tasked with protecting the park from the people was, however, unusually complex in this case. When conservation areas are created, existing areas of cultivation are often excised from their boundaries in a partial response to the accusation that it is the park that encroaches on existing farms rather than vice versa. In Lore Lindu, coffee stands were too scattered to be so excised; indeed, they coexisted with natural forest cover, in an example of the "ecology of coexistence" discussed above. Park rules dealt with this problem by registering already-planted coffee trees and allowing farmers to harvest their beans. Both the maintenance of existing stands and the planting of new ones, however, were forbidden. Farmers essentially had to treat their coffee as if it was wild, and over time, as the trees aged, berry harvesting would cease. Exclusion from the park was thus not total, as some farmers retained access but under severe restrictions on the activities they could engage in, and a time limit.

While the creators of the park felt that people had to be kept out of it, they were not prepared entirely to abandon those people to their fate. Rather, the park was further legitimated on the grounds that "integrated conservation and development programs" (ICDPs) would provide new, sustainable livelihoods for the excluded. Through the subsidies, incentives and technical inputs provided by these programmes, farmers would be assisted to mend their wayward practices and induced to stay out of the park. A wide array of external actors and experts, supported by tens of millions of dollars in funding, were mobilized behind these efforts. A programme run by the Asian Development Bank was supported through a gigantic $53 million loan and technical assistance package that was to cover the cost of training and equipment for park management, together with village "development" in the form of small-scale infrastructure, agriculture and health initiatives, and income-generating projects. Of this, $32 million had to be paid back, and a substantial part of that repayment was to be through conservation practices valorized as "environmental benefits". The monetary value of services such as improved water quality, carbon sequestration and biodiversity were imputed by the calculations of environmental economists to produce an Economic Internal Rate of Return of 18 per cent—even though no revenue for these "services" could actually be collected. The Asian Development Bank envisaged agricultural improvements and upgraded roads and health services as the pay-off that would reconcile villagers to the park. The Nature Conservancy ran a separate, much smaller programme that put grant money into small-scale income-generating schemes such as honey production. The main rationale of these schemes was to keep farmers busy, giving them neither need nor opportunity to encroach on the park.

From the Forest Department's point of view, the ICDPs succeeded in mobilizing the resources that turned the protected area from a paper park into a reality, complete with surveillance and enforcement. ICDPs also countered villagers' complaints that the park was excluding them from livelihood opportunities, since it was clear that large amounts of money were being spent on improvements outside the park boundaries. The park could thus be maintained as a zone of exclusion with a clear conscience. From the point of view of the villagers targeted by these interventions, however, they were miserable failures. The chequered history of ICDPs and buffer zone schemes in other sites has been widely documented.[23] In Lore Lindu, the agricultural schemes were not supported by land capabilities or by returns to labour and capital investments. Neither were they well thought out socially. The people most likely to suffer from being excluded from land and forest resources in the park—the landless or land-short—missed out on most of the ICDP interventions. The villagers' sense of exclusion was in fact exacerbated, not attenuated, by the involvement of large sums of foreign money and foreign experts, and by the discursive framing of the park as a global good. Villagers were convinced that rather abstract talk about the global good was merely an alibi for much more familiar and concrete benefit streams: officials would grab fat honoraria and kickbacks from project funds, and foreigners would grab the park's mineral and other resources. Powerful people, in short, would profit at their expense.

Ironically, however, the very ICDP programmes that brought villagers together (in good participatory fashion) to envision futures based on co-existence with an out-of-bounds park also provided platforms for the expression of common grievances. The many meetings with park planners and officials convinced villagers that their argument that the park boundaries were illegitimate impositions would not be recognized, nor would there be any serious response to the crisis of accessible farmland. Their rising sense that they were victims of injustice led them to connect with provincial-level NGOs that had already mounted campaigns against the exclusionary practice of the park management. A striking outcome saw villagers take possession of park land, defining their own zone of exclusion and autonomous management. Signs proclaimed the "Sovereign Domain of the Free Farmers' Forum", as villagers defied park managers and cleared new land along the road through the Dongi-Dongi Valley, an area well within the Lore Lindu boundaries. This occupation built on the earlier success of villagers elsewhere in the park who had avoided eviction by arguing that they were indigenous people, with a part to play in the park management system. The legitimation offered by the Free Farmers' Forum for the Dongi-Dongi occupation was rather different, however. The organization made a vague nod towards environmental stewardship, but it engaged with the park's conservation logic mainly in oppositional terms: farmers' need for land and livelihoods trumped the claims of nature, and of nature's protectors, who seemed to be oblivious to the suffering their exclusionary regime was inflicting.

A villager quoted by Li gave a poignant assessment of the injustice of exclusion when it interfered with the livelihood needs of a landless family, and the range of powers she confronted:

> We didn't have land. ... We were sharecropping rice. To get somewhere, you need to have land. So in 2000 we decided to try in the park. We came in with twenty people, to clear two hectares ... Then a party of forest guards came

by, fifteen of them. They came to my hut. I was getting the food ready for the workers. The workers ran to hide. They thought I had been arrested. So I gave the guards coffee and food. They stayed for one hour to eat. Then they said, "It's time for us to go. Do you know this is inside the park?" I said, "I know, but I need to eat." They asked me what I planned to plant. I told them candlenuts, cacao, durian, to replace the trees … They said, "Excuse us now" [i.e., they politely took their leave]. Then they started hacking at my hut and everything I had planted, cut it all to pieces, burned it down. I cried. I said, "God will see you. You have no pity." They just smiled. My cacao, coffee, chili peppers, they pulled it all up. I asked them, "Don't you eat chilis too? We are just going to grow crops, not take the land." They said, "You can't do that here, this place belongs to lots of nations." So I thought, does this land belong to Indonesia or to some other country? If it belongs to Indonesia, it belongs to me, too. Then they left. My workers came back and went right back to work, because I had already paid them …[24]

Self-exclusion in Community-Based Natural Resource Management

So far, we have considered conservation-based exclusions associated with projects designed at some distance from the communities affected, and by organizations whose agenda is framed in terms of the global common good. We now turn to the more counter-intuitive processes that engage people in excluding themselves from resource access. Projects that put decisions in the hands of groups envisaged by others (and sometimes by themselves) as "communities" implicate members of these "communities" in constraining their own land-based livelihood decisions and those of their neighbours. The idea behind CBNRM is that villagers will not just help implement exclusionary regimes, they will help to design them based on "local" criteria that take their place alongside state zoning schemes in determining appropriate land use. CBNRM is a discourse and a practice that has been widely adopted in a range of natural resource, institutional and political contexts.[25] Like many development-related concepts and approaches, CBNRM has its origins in a critique of mainstream practice. It stems from a concern that state-centric, centralized systems of forest, land, water, wildlife and fisheries management have failed to conserve natural resources and the environmental values associated with them, and have brought little benefit to rural people who depend on such resources. Proponents argue that ethnographically based understandings of the ways in which people manage their own resource base provide an alternative, more decentralized option. CBNRM resonates well with many ideas to which development agencies tend to subscribe, including emphases on decentralization and devolved governance, conservation, social benefits and empowerment of the rural poor. Common forms of CBNRM include community forestry, community-based fisheries management, community-based ecotourism, community-based wildlife management and community-based irrigation.

Community-based practices and discourses under the rubric of CBNRM take three main forms. First is the recognition and entrenchment, through processes of legitimation and institutionalization, of existing customary practices involving collective action in the field of natural resource management. Second is the more political process of articulating rights over resources that have come under threat of encroachment or exploitation by outsiders, whether those outsiders be neighbouring villages, large-scale corporate interests or state-led development projects. In particular, resistance to forest enclosures forms the galvanizing force for many CBNRM initiatives. Third is the programmatic expression of CBNRM in a range of initiatives by "big conservation" NGOs, small-scale civil society groups, governments and aid agencies. The involvement of institutions such as the World Bank in CBNRM is indicative of the ways in which once-radical ideas can be transformed into mainstream orthodoxy, and demonstrates the ease with which decentralized practice fits with neo-liberal governance agendas that include rolling back the state's role in rural development and environmental management. These different contexts of CBNRM also set it as a terrain of struggle, in which a shared terminology masks a range of contested meanings.

Despite the list of virtues associated with CBNRM, its capacity to effect win-win resolutions to the contentious politics of access and exclusion has proven to be limited. Community forestry, as one iteration of the CBNRM idea, provides a stark example of the conflict that arises when people contest their exclusion from the political forest and claim a role in managing their natural resource base.[26] In Thailand, villagers turned to community forestry as a way to counter the threats presented by loggers appropriating village resources and damaging their watershed quality during the 1980s. NGOs proposed a Community Forest Bill in the early 1990s with the aim of enhancing local rights to access and managing forest resources, and a long process of debate, drafting and counter-drafting led to numerous failed attempts to get the bill through Parliament. Ultimately a highly circumscribed version of the bill passed in late 2007 as one of the last acts of the military-installed government that had come to power the previous year. The bill devolves little authority over managing forest to rural people, as much of Thailand's remaining forest lies within protected areas defined in the bill as off limits to community management. The inclusive discourse of the bill's preamble is belied by the rigid exclusions set out in the finer print. In Indonesia, similarly, decades of donor-sponsored initiatives in community-based or "social" forestry have had remarkably little impact in shifting the exclusions of the political forest.[27] In the Philippines, in contrast, the community forest approach has become an important vehicle for legitimating the presence of farmers in the forest and formalizing their tenurial rights, as we noted in Chapter 2.

While CBNRM has achieved some success in devolving management to village-based decision makers whom farmers and other rural residents may respect and can influence, it also has its own exclusionary effects. The first of these is territorial. CBNRM programmes usually begin with an exercise in mapping the territory claimed by the "community" in question, a process that often involves drawing firm lines where outsiders see customary practice as vague, and where territorial rights are not rigidly specified. Unclear boundaries between villages that share de facto access to a river or lake or adjacent patch of forest are an obvious example. Of course, these seeming ambiguities also may be based on customary practice in which "shades of grey" allow overlapping access for grazing, forest product collection, fishing and other use of the local commons. The second exclusion is social, when "the community" is defined in a way that makes migrants or minorities second class, ineligible to participate in the new management system. As customary and new rules are formulated, migrants and minorities may find that the de facto rights they enjoyed come up for scrutiny and are marginalized or rejected outright. People who have moved or been moved multiple times as a result of war, resettlement schemes or evictions are not usually regarded as good targets for CBNRM, which favours cohesive communities with customary territories of long standing. They are especially vulnerable, as the "communities" around them take advantage of CBNRM-legitimized resource claims to tighten up their exclusionary management regimes.

Conflicts over exclusion are intensified when CBNRM initiatives are hooked onto income-generating schemes that increase the value of the resource in question. In Kalimantan, the Indonesian portion of Borneo, for example, villagers saw "community" resource maps as tools in the struggle to demarcate forest boundaries vis-à-vis neighbouring villages, a matter that became urgent when new rules permitted village leaders to enter directly into contracts with outside entrepreneurs for timber extraction. In Laos, an initiative to consolidate village "management" of a wetland in order to increase the commercial offtake of fish had the effect of excluding members of neighbouring villages who had previously enjoyed the customary right to access the backswamp for their own livelihood needs.[28] More often, however, the CBNRM schemes promoted by outsiders (national and transnational donors and NGOs) tend to be hostile to market engagement. Despite the talk of "sustainable livelihoods", their main objective is conservation. In part, this is a response to the "ambient environmentalism" that requires resource claims to be legitimated in environmental terms. The result can be to paint villagers into a corner in which they forever remain marked as culturally distinct, environmental subjects rather than equal actors in a marketized rural economy (see also Chapters 2 and 7).[29] Referring to the case of the Karen minority in Thailand, Andrew Walker argues that their long-standing experience of production for markets is overlooked, and their farming practices—indeed their very identities—are "arborealized", or made one with the forest.[30] Ethnic minority farmers (often described as indigenous people), and NGOs and academics advocating with them or on their behalf, may argue for their right to "exist in nature" as a strategy to sustain their access to land, or they may believe that this is the proper way to live. Either way, the result is to misconstrue their history and limit their options for the future, a point we take up further in Chapter 7.[31]

CBNRM in Cambodia

The Khmer term for "community" is *sahakom*. The word is rarely heard in farmers' vernacular usage to describe village-level activities or collectivities, in part because it is close to the term *sahakor*, or cooperative, and this immediately conjures up the forced collectivization imposed during the Khmer Rouge regime.[32] Yet we see legislation on community forestry, community fisheries, officially designated "Farmer Water User Communities" and numerous NGO-inspired community-based natural resource management programmes employing the term *sahakom*. Use of the term raises questions about whose collective action, and whose interests, these schemes represent.

A close reading of some well-documented cases provides insights into the legitimating framework for CBNRM programmes in Cambodia and elsewhere. In particular, it shows how self-exclusion can be used defensively when resources are under threat.[33] Consider the following narratives:

> In October 2003, villagers discovered poachers cutting timber from their recently formed community forest in Toupcheang, Koh Kong province, Cambodia. There had been a longstanding problem with armed outsiders entering the forests in their community and logging valuable luxury-quality timber … the poachers cut timber virtually uncontested, saying 'we have the guns, you don't …' [F]rustrated by tensions with poachers in their own forests, the Koh Kong villagers began to consider how community forestry might benefit them … They began negotiations with a concessionaire company … Together, the company and villagers developed rules and regulations for the forests and with the help of the provincial forestry officers were able to convince the district governor to approve them.[34]

> Since Cambodia is a strongly hierarchical social context, having high-level
> political support for NRM activities is essential ... Consider the fisheries
> reform. In 2000, the prime minister of Cambodia visited the provinces and
> heard about conflicts between fishers and fishing lot owners ... It is perceived
> by many government officials that villagers have a low capacity or limited skills
> and experiences for resource management ... The challenge is to break down
> negative perceptions while getting higher-level officials to support CBNRM
> process. ...[35]

These excerpts reflect a two-pronged approach taken by CBNRM programmes in Cambodia. The catalyst for many initiatives is an encroachment on forest, fish, water or land resources by powerful outside interests. External intermediaries work with and on behalf of villagers to advocate provincial government intervention to stop the encroachment and the establishment of clear village boundaries and hence resource rights. Part of this negotiation involves demonstrations that villagers have their own established customary management system, have the capacity to improve management, and have the conservation of natural resources in mind. Typically this will involve work at the local level to establish a set of rules about who can access what resources, prioritizing extraction of resources for subsistence rather than commercial purposes, and zoning within village boundaries. A ceiling is sometimes placed on landholdings for cash crops. In order for community fisheries and community forests to be recognized officially, they have to go through a prescribed set of formal procedures that delimit, zone and set rules for the use of resources, monitored by provincial officials and approved at the ministry level. They also must subscribe to the relevant sub-decrees, for example, a provision in the Community Forestry sub-decree that stipulates a moratorium on trade or barter in forest products for five years after ministerial approval of a community forest. In other words, the exclusion of external players from village territories and resource exploitation requires legitimation through a process of self-exclusion, with the whole operation framed in terms of an ambient conservation agenda.

The prevalence of CBNRM initiatives in Cambodia can be attributed in part to the country's distinctive history. The dramatic social and political upheavals of the Khmer Rouge years sometimes overshadowed, but also provide an important context for, the emergent regime of land access. The country still has abundant natural resources, and an unusually large proportion of the population depends directly on these resources for everyday livelihood purposes. In recent decades, however, rapacious and inequitable exploitation of natural resources has put many rural people's livelihoods under threat in a manner that is closely linked to environmental quality and poor environmental stewardship. Exploitation takes the form of land-grabbing in upland areas for plantations, intensive logging, gem mining, sand extraction for construction and for export to Singapore for landfill, destructive fishing practices, and, increasingly, hydropower development. CBNRM initiatives, strongly promoted by international donors and NGOs, have entered this scenario as attempts to maintain both the resources themselves and access to them by the rural poor.

A survey in 2002 showed that there were 237 community forestry and 162 community fisheries projects in Cambodia.[36] Most of these were run by NGOs. However, CBNRM has also been taken up by government agencies and been institutionalized in government policies. A Community Fisheries sub-decree enacted in 2001 provides for government-recognized community management of fisheries in the Tonle Sap Lake (Great Lake), including provision for the transfer of 58 per cent of the large fishing lots from their long-established private concession basis to community-based management. These fishing lots are leased by wealthy individuals with their own locally hired enforcers.[37] At the provincial level, agreements have been negotiated to curtail some of the large, central-government-granted private logging and plantation concessions that had encroached on villagers' agricultural and forest-based livelihoods.

An iconic site for CBNRM in Cambodia is the northeastern province of Ratanakiri.[38] Most people living in the province are non-Khmer ethnic minorities from nine different language groups. From the late 1990s, Ratanakiri, with its mainly intact forests and rich volcanic soils, emerged as a resource frontier. The central government granted timber and oil palm or rubber plantation concessions to domestic and foreign companies with little or no consultation with the provincial government. Indeed, so poorly coordinated were the concessions and the overlapping establishment of protected areas that at one stage it was calculated that 115 per cent of the province's land area had been allocated to these uses! The first that many of the shifting cultivators who form the majority of people indigenous to Ratanakiri knew of the concessions was when loggers protected by armed guards arrived to clear forestland and fence off the concession areas. The areas enclosed included villagers' swidden plots, fallows, and sometimes even paddy lands, as well as large areas of forest from which villagers collected non-timber forest products (NTFPs).

At the same time, the national governance framework was being taken in the direction of decentralized management through the Seila programme, which was established in the 1990s to rebuild local governance capacity and supported by a United Nations Development Program initiative called the Cambodian Area Regeneration and Rehabilitation Project. From the mid-1990s, with the intervention of NGOs working in Ratanakiri, the Seila programme took on the specific concerns of indigenous people facing encroachment on their lands and natural resource base. Villagers in Somthom commune went through a cartographic exercise of identifying and marking the boundaries of the resources that they used in a process of participatory mapping. They documented their pre-existing ("customary") uses and ways of managing those resources. Their maps zoned forestland into fallows, spirit forests, NTFP collection areas and watershed forests, and also identified agricultural land and water bodies. The main purpose of the maps, which were computerized and presented to commune, district and provincial governments, was to demonstrate the problem of concessions that overlapped village land, a problem that received a sympathetic hearing at provincial level (in part because all the revenues from concessions go directly to the centre). But a further purpose of the process—the second prong of its ambient environmentalism—was to establish natural resource management committees in each village to formulate and implement rules and regulations both within communities and vis-à-vis external parties, including neighbouring villages.

Of particular interest in this case, and widespread in CBNRM initiatives as we noted earlier, was the rule devised to limit the area of land that each household could convert to perennial cash crops, with a ceiling of 2 hectares per household. Implicitly, for CBNRM to thrive, market powers—especially the attraction of lucrative crops that might draw villagers into a boom mentality (see Chapter 4)—needed to be kept at bay. The limits on farm size were in part based on the desire to prevent land accumulation and alienation among villagers, keeping them more or less equal; they were also based on an argument about cultural difference. Specifically, CBNRM advocates attempting to persuade authorities to strengthen villagers' land rights often stress that indigenous minority villagers are different from the exploitative outsiders encroaching on their resource base and are willing to limit their agricultural ventures accordingly.

The Ratanakiri example confirms the general pattern we described above. The impetus for CBNRM was defensive, as villagers and their advocates attempted to counter their actual or threatened exclusion when the central government allocated concessions overlapping their land. To achieve this goal, villagers were led to support a territorialized approach to resource regulation that expanded to elicit their participation in devising a regime to limit their own practices and exclude themselves from accessing resources they used previously. Often, CBNRM initiatives also led villagers to exclude other "ordinary" people, usually people they defined as outsiders, a practice that was especially problematic in circumstances where pre-existing access to resources was non-territorially based. A significant example in Cambodia is the case of seasonal migrants who move to the edge of Tonle Sap for fishing during the agricultural off season. CBNRM under the community fisheries programme has encouraged permanent settlements to draw boundaries that exclude those who previously accessed fish on a seasonal basis.[39]

Corporate Conservation and Eco-governmentality

In the discussion on protected areas, we saw that, contrary to a narrative that poses environmental protection as the antithesis of development, protected areas share two important features with large-scale development projects such as dams: they exclude people from access to land, and they legitimate exclusion in the name of a greater good. More surprising still is the insertion of conservation concerns into corporate development projects that package up market-oriented and conservation rationales to present themselves as a "win-win" outcome for many actors, although they are often disastrous for the people displaced. This approach to project design spreads and intensifies exclusionary effects to encompass not only the people directly displaced by the projects themselves, but the people using resources in the surrounding areas who are subject to the intensified surveillance that accompanies the spotlight effects of a large project. There may be moves to expand conservation zones by resettling people not directly in the path of the project, in order to compensate for its environmental effects. Ambient environmentalism, in short, may result in a corporate project design that signs away many people's access to land and livelihoods without consultation and with remarkably little public debate.

The ambient environmentalism of large projects has three sources. First, the era of corporate social responsibility has brought with it the "triple bottom line", in which corporate success is measured in terms of social and environmental as well as financial outcomes. While the motivations behind these moves are often debated, the effects of schemes implemented in pursuit of the triple bottom line are more important than the goals that drive them. Their exclusionary effects are expansive and may be located far from the principal site of commercial interest. Debt-for-nature swaps are one example: a New England power station subsidizes the intensified management and policing of rainforest reserves in Central America.[40] In this case, the intervention is seen as compensatory. Other initiatives involve framing a commercial investment in environmental terms, as when, for example, tree plantations for pulp and paper or for biofuel are represented as carbon sequestration and fossil fuel replacements respectively.[41]

Second, what Peet and Watts refer to as "market triumphalism" has seen market-based solutions applied to conservation problems.[42] In this case, commodification of nature is presented as a governance solution to market failure. One application of this logic is the notion that exclusive property rights are necessary to solve the "tragedy of the commons", enveloping latter-day enclosures in an environmental rationale.[43] More generally, when the economic incentives are right, so the argument goes, conservation will follow.

Third, what Michael Goldman terms "eco-governmentality"[44] has given the development banks a new mission as guardians of the "global commons". In pursuit of this mission, the World Bank requires its developing country clients to implement rafts of environmental legislation as a condition for receiving loans. Environmental conditionality has also crept into projects that are essentially commercial in nature, especially where development banks fund the government's stake in an enterprise and cover the sovereign risk guarantees that give private banks and other investors the confidence to proceed. This fusion of the public and private realms also leads to project-specific conditions such as "offset" conservation stipulations that are supposed to protect an area equivalent to that destroyed by the project itself, a prominent feature of the Nam Theun 2 project in central Laos.

Nam Theun 2

The Nam Theun 2 hydroelectric project is a large and extremely complex scheme, with a history and scope that take its significance well beyond the immediate impact area. The project has been the object of critical academic study,[45] of voluminous project documentation from nearly two decades of appraisal and pre-construction monitoring,[46] and of extensive NGO critiques.[47] While the negative environmental impacts of hydropower projects are widely documented, proponents of Nam Theun 2 have presented it as positive for the environment. Project documents suggest that without hydropower, Laos would have no choice but to increase timber exports. Project revenues, it is alleged, will support the protection of forest in the dam's catchment area. The project in its World Bank-underwritten configuration was advertised as the only alternative to a much more damaging version that would otherwise have been funded and built by Chinese companies. The project's environmentally benign image received a further boost from positive evaluations of hydropower as a non-fossil-fuel-based energy source in the age of climate change. The recasting of Nam Theun 2 as an environmental project, and its exclusionary implications, are our main focus here.

The main dam in the Nam Theun 2 project is a 50-metre-high barrage on the Theun River, which is the fourth-largest Mekong tributary measured in flow volume. Nam Theun 2 is a relatively small impoundment structure, but it is the largest hydropower project on a Mekong tributary.[48] The scale of the power project derives from the diversion of the river off the Nakai Plateau, through a tunnel down an escarpment into a different river basin, giving the turbines a head of water of about 350 metres. The dam has cost over US$1.5 billion; at the time the project was planned, this was more than the annual gross domestic product of Laos. The power station's electricity output is almost entirely for export to Thailand.

When it was first appraised in the late 1980s, Nam Theun 2 was to be a publicly financed and government-utility-owned infrastructure project. Since the mid-1990s, however, it has become a corporate enclave project, which means that its financing, ownership and management are quite separate from the national economy in which it is located. Nam Theun Power Company is 75 per cent owned by international private investors. Nevertheless, without World Bank financing to assist the Lao government in investing its equity share and to provide sovereign risk guarantees, the project would not have gone ahead. The World Bank has thus retained important leverage in the project design.

The reservoir behind the dam flooded 450 square kilometres of forest, agricultural land, and settlements of a number of ethnic minorities on the Nakai Plateau. Until the mid-1990s, these people practised a combination of paddy cultivation, swidden farming and forest product collection. The villages of those affected had been located on the plateau for many generations, except for a period when the area was bombed during the Second Indochina War. In the early 1990s, few people on the plateau knew of the already well-advanced plans for the dam.[49] By the late 1990s, the area had been transformed as advanced clear-felling took away a principal source of livelihood and as preparations got under way for the project. Things did not, however, proceed smoothly. The dam was delayed by a complex, drawn-out and acrimonious process of opposition; below-standard appraisals; the reconfiguration of the project's corporate structure in the wake of the 1997 financial crisis; studies and restudies in response to critiques of resettlement and environmental mitigation and management plans; and consultations and public hearings at the village level, in Vientiane, and in several world capitals. As a result, those directly affected by the project remained in limbo about the project's fate for more than a decade.[50]

The project's impacts are both direct and indirect. When the dam was sealed in 2008, some 6,000 people lost their land and surrounding forest resources to the reservoir, and they have been resettled. The initial plan was to resettle people in the lowland valley below the Nakai escarpment, but consultations revealed that most preferred to stay on the plateau, close to the edge of the reservoir. Beyond the dam site and the inundated area, there are significant downstream impacts on two rivers. The Theun River will see its flow greatly reduced, since nearly all the water during the dry season and a significant proportion of wet season flows will be diverted into the Nam Kathang/Xe Bang Fai River systems. These recipient channels, meanwhile, will need to accommodate run-off several times larger than their natural flow regime, creating severe impacts on fisheries and enhanced flood risk in one of the most fecund tributary systems of the Mekong Basin, in a river valley that is already subject to extensive flooding.

Of more immediate significance for the present discussion, however, is the indirect impact upstream of the reservoir. Before the World Bank became involved with Nam Theun 2, the Global Environment Facility was considering funding a very large National Biodiversity Conservation Area linked to a contiguous protected area across the border in Vietnam, in the upper catchment lands draining into the Theun River. This remote part of the Annamite mountain range became the subject of global attention following the scientific discovery in 1992 of the Vu Quang ox, the first new bovine genus to be discovered anywhere in the world since the early 20th century. Equally important to the immense biodiversity significance of the upper catchment area, however, is the protective value of the watershed forest. The dam's reservoir is located on a plateau and, being only 13 metres in depth at its deepest point at the maximum level of inundation, is extremely vulnerable to siltation. Some 98 per cent of the upper catchment remains under forest cover. As early as 1996, project proponents began to argue that the project would have an incentive to avoid siltation-inducing deforestation and would therefore be willing to finance the policing of a protected area that would be more than a paper park. This was another way in which Nam Theun 2 would be good for the environment.[51]

The upper catchment area marked for conservation is sparsely settled, but it is not wilderness. Around 6,000 people in 31 villages inhabit the area in what were, until the project was started, extremely remote forest-based communities. The concession agreement with the Nam Theun Power Company stipulates that the dam provide US$1 million per annum over the 30-year concession period to manage the protected area through an integrated conservation and development programme, which would involve a combination of livelihood support projects and restrictions on land use. A great deal of emphasis is put on participatory land use planning and other consultation-intensive attempts to refashion livelihoods along conservationist lines. There is little doubt that clean water programmes, schools, health centres and other facilities will improve basic material conditions for those living in the protected area. There is also little doubt that these people will be restricted in their livelihood activities, with constraints on fallow cycles and strict application of rules prohibiting hunting and gathering of tradable forest products.

Moving to an area even more remote from the project site, we find another category of exclusion that is—ironically—even more intensive. A further 48 villages near the edge of the protected area have been classified as being located in the Project Impact Zone (PIZ). Unlike the villages in the protected area itself, where a limited degree of resource access is tolerated, the project seeks to keep villagers from the PIZ out of the watershed area altogether. In these villages, too, an ICDP seeks alternative livelihood development as a compensatory measure in recognition of the loss of resources due to exclusion from the National Biodiversity Conservation Area.

We conclude with an example of the way in which the environmental goals of international institutions can run ahead of reality. A relatively late entrant into the Nam Theun 2 project, the Asian Development Bank has applied its own safeguard measures to ensure that the net environmental effect of the investment is neutral or positive. One of the ADB's contractual stipulations was that an area of forest equivalent to that flooded by the reservoir (measured at 28,000 hectares) should be planted within the upper watershed area. Only later did the bank and project managers realize that, in an area in which 98 per cent of the land remains forested and the remainder is fallow land in active swidden rotation, there is virtually no scope for increasing forest area. The ADB has worked around this by indicating that it will support afforestation in "abandoned" fallows in nearby districts and provinces, effectively taking land out of the swidden cycle of farmers otherwise unconnected to Nam Theun 2. Quite how it will meet this objective is yet to be seen, but the plan would seem to embroil the ADB in intensifying and funding the Lao government's highly controversial programme of resettling upland-dwelling people away from their lands and undermining their swidden-based livelihoods.[52]

Conclusion

This chapter has examined the pervasive and increasing role of what we have called ambient environmentalism in excluding large numbers of Southeast Asians from access to land. Conservation initiatives are typically based on well-intentioned and often creative attempts to serve the common good and to protect not only natural landscapes and organisms, but also the resource base on which rural people depend. As we argue throughout this book, however, all land uses, including conservation, require exclusion, and someone pays the price. Often that person is a relatively powerless villager, although villagers also exercise agency in relation to conservation as they pursue their own agendas, as we have seen.

In this final section, we summarize the different powers deployed to exclude people from access to land on the grounds of conservation. The common good of current and future generations that is inherent in conservation and sustainability discourses provides a pervasive moral argument for exclusion. States and international development agencies mobilize constituencies through discursive legitimation at a number of levels. These constituencies include the growing urban populations of Bangkok, Manila, Jakarta and other centres who subscribe to the idea of a wild nature that should be protected from people, unless they plan to use it lightly, for purposes of recreation. They include the state agencies and consulting firms that hitch the virtues of conservation to revenue streams, as donor funds flow into their coffers. They include corporate actors whose projects—otherwise portrayed as destructive of natural values—can be recast as environmentally enhancing. And, to an extent, they include people transformed into environmental subjects in "community" projects that reframe a defence of livelihoods as a search for sustainability, in which self-exclusion plays a part.

While legitimation is a sine qua non for conservation-inspired exclusion, it is regulatory power that governs where and how people live and farm. It is not only state actors, however, who wield this power. National parks and wildlife reserves are the creatures of states, but international conservation agencies and development banks facilitate and often finance the implementation of the broad-scale regulation inherent in protected area establishment and management. A more broadly ambient "eco-governmentality" employs global commons discourses to regulate land use through influence on national laws as well as through schemes such as certification that govern how farming is to be done, and use market-based incentives to encourage or enforce compliance. CBNRM requires "communities" to regulate themselves by restraining their use of land and natural resources, and often by limiting their access to potentially lucrative livelihoods. In some cases, the exclusionary effects of CBNRM are rather similar to those of the top-down projects they are intended to replace.

We have not highlighted force as a power of exclusion associated with conservation, and indeed this is not a dominant or pervasive power in this case, although it comes into play when people are evicted and their crops destroyed, as in our example from Sulawesi. Force also enters into conflicts between "communities" struggling over resources and conservation values. In one egregious case in northern Thailand, similar to an instance that comes up in our discussion of collective mobilization in Chapter 7, farmers associated with the Dhammanat Foundation in Chiang Mai province argued that minority upland farmers were damaging the environment, and threatened a group of Hmong farmers with violence. They burned effigies of their academic and NGO supporters. In another case, lowland farmers in Nan province burned the houses and lychee orchards of upland Hmong. Lohmann describes such actions as "forest cleansing", in a direct analogy to the violent racist practice of ethnic cleansing.[53] Observers of northern Thailand argue that in the relatively rare cases where evictions from forest have taken place, tensions between majority and minority communities have been important factors prodding state authorities to take action.[54]

Finally, the power of the market enters conservation through various channels. Generically, the attribution of market values for environmental goods gives a neo-liberal rationale for excluding those whose livelihood practices are deemed to reduce those market values. There are schemes to impute market values for environmental services, and pay people to deliver those services by compensating them for the value of the current land and resource uses they have to give up. Two contexts in Southeast Asia where these kinds of schemes have been mooted are upper watershed protection and climate change pre-emption, the latter spurred on by global provisions for the Reducing Emissions from Deforestation and Forest Degradation initiative. The lack of property rights, or even citizenship rights, is one among a number of institutional obstacles for market-based conservation programmes to take hold, since it is difficult to pay people for loss of access to resources that are not officially recognized as theirs in the first place. Nevertheless, the market has been brought in through corporate involvement in conservation, where part of the economic fruits of commercial resource development projects are diverted to conservation in packaged arrangements such as those found in Nam Theun 2. Everywhere one looks in Southeast Asia, conservation is in the air.

Notes

1. McCarthy 2006: 5.
2. Peluso and Vandergeest 2001.
3. See Li 2007c. See also Geisler 2003, Peluso and Vandergeest 2001.
4. Estimates of the number of people living on land classified as public forest in various Southeast Asian countries in 1992 were as follows: Indonesia 40 million–65 million, Philippines 24 million, Thailand 14 million–16 million. Source: Lynch and Talbott 1995: 22.
5. Goldman 2004.
6. Geisler 2003: 71.
7. Agrawal and Redford 2007: 7.
8. Geisler 2003: 71.
9. Peluso and Vandergeest 2001.
10. Boomgaard 1988.
11. Protected Areas and Development Partnership 2003: 13, 17.
12. Geisler 2003: 69.
13. Wong 2006.

14. Hirsch and Wyatt 2004.
15. Geisler 2003: 70.
16. Walker and Farrelly 2008.
17. See New Mandala, http://rspas.anu.edu.au/rmap/newmandala/2008/09/23/information-about-village-evictions-in-northern-thailand/.
18. Brockington, Igoe and Schmidt-Soltau 2006: 251.
19. Redford and Fearn 2007.
20. Agrawal and Redford 2007.
21. Fisher and Maginnis 2005.
22. Unless otherwise indicated, this section is based on Li 2007c. See especially Chapter 4.
23. Hughes and Flintan 2001: 7; Brandon and Wells 1992.
24. Li 2007c: 142–3.
25. For a nuanced discussion of the various agendas and viewpoints on CBNRM, see the excellent collection of case studies and arguments in Brosius, Tsing and Zerner 2005. For a rich set of case studies on CBNRM in action, see Tyler 2006.
26. Hirsch 1998b.
27. Li 2007b.
28. Tubtim and Hirsch 2005.
29. Agrawal 2005.
30. Walker 2004.
31. Walker 2001. See also Li 2002a for a critique of the application of simple and universalistic legal approaches to CBNRM and its assumption that distinctive customary "tribal" practice represents the main arena for empowerment vis-à-vis the state.
32. The term *phum*, or village, is much more common. The authors are grateful to Ken Serey Rotha for this insight.
33. For a range of case studies that reflect on CBNRM through action research, see Tyler 2006.
34. Kamnap and Ramony 2006: 209–10.
35. Nong and Marschke 2006: 162–3.
36. McKenny and Tola 2002.
37. Degen and Thuok 2000.
38. For a detailed account of the history and challenges of CBNRM in northeastern Cambodia, and specifically in the commune of Somthom, see John and Phalla 2006. The following material is derived from their study.
39. Sitthirith 2006.
40. Vidal 2008 calls these practices "eco-colonial", http://www.guardian.co.uk/environment/2008/feb/13/conservation.
41. Ibid.
42. Peet and Watts 1993.
43. For a critique of market-based approaches to conservation and their social justice implications, see Zerner 2000.
44. Goldman 2004.
45. Goldman 2005, Hirsch 2002, Wyatt 2004.

46. See the World Bank website on Nam Theun 2, www.worldbank.org.
47. See the websites of International Rivers (www.irn.org) and TERRA (www.terraper. org).
48. As of 2010, there were no dams on the Mekong mainstream in the four lower countries of the Mekong River Basin, although there were several on the mainstream in China with generating capacities larger than that of Nam Theun 2.
49. See the critique of early studies in Hirsch 1991.
50. See Hirsch 2002 for a discussion of the extraordinary extent to which Nam Theun 2 became disarticulated from Lao debates and became embroiled within a global amalgam of interests around environment, development and social justice concerns.
51. David Iverach, in the SBS Dateline documentary film *Dam Destiny*, screened 9 Aug. 1996.
52. Baird and Shoemaker 2007, Vandergeest 2003a.
53. Lohmann 1999.
54. Walker and Farrelly 2008.

4

Volatile Exclusions: Crop Booms and Their Fallout

This chapter addresses the transformations of land access that follow from the occurrence of booms in the production of crops such as coffee, oil palm and shrimp. Crop booms have brought the promise and reality of prosperity to millions of Southeast Asians but have also resulted in the exclusion of large numbers of people from access to land through the processes that accompany and follow in the wake of a boom. When booms begin, the prospect of enrichment often stimulates large-scale migration to boom areas and the transfer of land to the newcomers. The multi-year time horizons of many boom crops encourage producers to make formal and individualized claims to land previously held under more flexible understandings. And the crashes that so frequently follow booms magnify the more routine processes of accumulation that see some people prosper from a boom, while others lose their land. The dynamics of exclusion associated with booms pose dilemmas for both state actors and producers. To officials, boom crops promise large-scale foreign exchange revenues and the modernization and development of the countryside. But officials also worry that the expansion of agricultural land will conflict with conservation goals; that conversion to export crops will jeopardize national food security; and that farmers who embrace monocropping will be left with no livelihood should the crop fail. This last worry is shared by farmers who have to decide how much land, labour and capital to commit to the boom. Growing a boom crop can generate many times more income than planting a field to rice. But it is also a gamble, involving what can be a hard-to-reverse commitment to a fickle crop. Disease and/or falling prices have wrecked the dreams of boom crop producers across Southeast Asia.

Our understanding of crop booms has three main features. First, booms involve the rapid conversion of large areas of land to monocropped (or nearly monocropped) production. The key triggers for this process are rising crop prices, the introduction of new growing techniques, and various kinds of state support. The Southeast Asian shrimp aquaculture boom of the 1980s, for example, was sparked by rapidly rising demand (especially from Japan), new hatchery technologies and government promotion. Second, booms involve transformations in land use with time horizons of more than a year. While maize production expanded very rapidly in upland Thailand after World War II, the fact that fields planted to maize could easily be switched to something else in the next growing season meant that the long-term implications for land relations of the specific choice of crop were limited. Tree crops such as rubber, coffee and cocoa, however, take years to come to maturity and begin producing, and this often prompts boom participants to try to establish new and permanent forms of exclusion. Shrimp farming also has this characteristic because of the need to construct ponds, and because salinization makes it difficult to convert shrimp farms back to other agricultural uses. Third, crop booms with major fallout for land relations usually involve three important sets of actors. Smallholders are at the heart of many crop booms. States play a critical role by supporting migration and providing subsidies, loans, agricultural extension and infrastructure. Agribusinesses (national and foreign) are also important for their roles in establishing plantations, supplying key production technology, engaging smallholders in contract schemes, and setting up downstream processing and marketing capability.

For the sake of simplicity we label the crops we discuss here "boom crops", but we do not suggest that their production takes place *only* as part of booms. Nor is it necessary that any of these crops be monocropped, or that their growers devote all of their land to them. As we discuss in Chapter 6, for instance, rubber is sometimes a boom crop with transformative effects (as it is currently in Laos), but it has also been integrated into swidden practices without causing a major transformation in land relations.[1] The relationship between the "crop" and the "boom" is thus an elective affinity rather than an ironclad regularity. In our treatment of the "crop" side, too, we wish to avoid "crop essentialism", the notion that the biological characteristics of a crop determine the organization of its production. As the discussion in this chapter makes clear, each of our boom crops has been produced under too wide a range of organizational frameworks for any simple biological determinism to be at work.[2] We do show, however, that the characteristics of crops, and of their downstream processing facilities (which may have economies of scale), are consequential for the way they are grown.

Crop booms have a long history in Southeast Asia. As Tania Li writes elsewhere, during the 19th century colonial observers "were repeatedly surprised, and horrified, by the willingness of farmers [...] to take up production of 'boom' commodities such as rubber, coffee, and coconut, and their rapid slide into debt resulting in the mortgage and sale of land" (see also Chapter 6).[3] In the late 20th and early 21st centuries, six main boom crops can be identified in Southeast Asia: cocoa, coffee, fast-growing trees such as acacia and eucalyptus, oil palm, rubber, and shrimp. We discuss the Sulawesi cocoa boom in Chapter 6. In this chapter, we take up three other crops and study their expansion in specific contexts: oil palm (in the Malaysian state of Sarawak), shrimp (in Thailand) and coffee (in the Central Highlands of Vietnam). Important similarities link these crops. All three are grown mostly for export and are important sources of foreign exchange for the producing countries. Each is associated with visions of prosperity and modernity. Each has expanded at the expense of forests, and each has seen extensive conflict. Oil palm, shrimp and coffee also demonstrate, however, the range of ways in which powers of exclusion have been mobilized in the context of booms.

Our oil palm study focuses on Sarawak, one of the front lines of the oil palm boom. Sarawak state agencies began seriously to promote the crop in the 1970s and devised a series of (not particularly successful) mechanisms for bringing together land, capital and labour. In the mid-1990s a new scheme was introduced that tried, in neo-liberal fashion, to promote joint ventures between private corporations and government agencies acting as trustees for "native" communities (which would provide land and labour). While this framework allows oil palm to be legitimated as a means for lifting natives out of poverty, in reality private plantations employing migrant labour from Indonesia and elsewhere are responsible for most of Sarawak's production. There is ongoing contention over the land on which these plantations operate, much of which is subject to native claims that are not recognized by the state. Nevertheless, native smallholders are keen to plant oil palm when they have the opportunity. Indeed, it is one of the ironies of Sarawak's boom that while some state actors are promoting the smallholder side of the boom, others are discouraging it.

Shrimp farming swept the coasts of Southeast Asia in the mid-1980s. Thailand was among the first countries to jump on the bandwagon and was the most successful, dominating Asian farmed shrimp production through the 1990s. State agencies, international donors, and Thai and foreign agribusinesses were critical in spreading the technical innovations that drove the boom. A primarily owner-operated shrimp farming sector comprising quite small farms (usually 1–2 hectares) appeared very quickly. Of the boom crops we examine in this chapter, shrimp farming holds out the most spectacular promise of financial gain, and huge numbers of people—farmers and fishers, but also urban professionals—have taken the plunge. Shrimp, however, is also the riskiest of Southeast Asia's boom crops, and Thai shrimp farming has been highly unstable. A series of production crashes has kept shrimp production peripatetic, as diseased and polluted ponds have been abandoned and new areas opened up. We identify four key exclusionary processes that have characterized the sector: the enclosure and conversion of mangrove forests, the effort to ban inland shrimp farming, the use of force by shrimp farmers to deter theft, and production crashes.

The coffee boom in the Central Highlands of Vietnam, finally, began abruptly around 1993 as world coffee prices soared while agricultural production was being devolved from the collective to the household level. The boom transformed landscapes and livelihoods across the highlands. Sponsored mass migration to the area from the lowlands had been going on since the end of the Vietnam War, and the boom saw both the enthusiastic participation of these earlier arrivals and an increase in migration. Much of the land migrants farmed had been acquired, by various means, from upland minority groups, a process fraught with exclusionary dynamics. Uplanders also took part in the boom, and moving from more diverse cropping strategies to a focus on coffee brought great changes to the organization not just of production but of land access and social life more broadly. When the crash came, it was driven not by pollution and disease but by a collapse in export prices. The shakeout that followed contributed to ethno-territorial conflict between uplanders and migrants in the Central Highlands, a topic we take up in Chapter 7.

Palmed Off: Oil Palm in Malaysia

Anyone who has spent time driving through rural Malaysia is familiar with the orderly landscape of monocropped oil palm. Stretching off on both sides of the road, the rows of trees could be anywhere. Often the terrain and former settlement sites have been bulldozed, giving no sense of what was there before the palm. Homogenous landscapes have always been a feature of industrial agriculture, and there is nothing new about oil palm in this regard. Nevertheless, being confronted with endless vistas of these trees can inspire strong emotions. To some, the massed ranks of trees evoke prosperity, rural people lifted out of poverty, and a rational, modern countryside. To others, what comes to mind is what has been concealed or obliterated: workers abused or poisoned with pesticides, water polluted, land grabbed, people evicted, forests cut down (and often burned, contributing to the infamous hazes that periodically blanket much of Southeast Asia), and orang utan and other wildlife threatened with extinction.[4] NGOs such as Greenpeace, indeed, have been waging a high-profile campaign against oil palm expansion in Indonesia and Malaysia and against companies, such as Nestlé, that allegedly make use of "unsustainable" palm oil.[5]

Whatever one's views on the crop, however, oil palm looks like the future. During the 1990s and 2000s, the two main supply-side developments in Southeast Asia were the extension of the crop from Peninsular Malaysia to Sumatra and Borneo, and Indonesia's rapid drive to pass Malaysia as the world's top producer (see Figure 4.1). Astonishingly, between 1990 and 2007 nearly 8 million hectares of land were converted to oil palm in Malaysia and Indonesia.[6] As early as 1997, oil palm covered more than one-tenth of all the land in Sabah, and nearly three-quarters of the state's agricultural land.[7] The success of Malaysia's and Indonesia's drive is reflected in the fact that by 2004, the two countries produced 85 per cent of the world's palm oil and accounted for 89 per cent of palm oil exports.[8] Government statements suggest that much more is to come. In the mid-2000s, plans in Indonesia envisaged expanding West Kalimantan's 2002 planted area of 338,000 hectares to 3.2 million, pushing East Kalimantan from 70,000 to 2.1 million, and lifting West Papuan production from 58,000 to almost 3 million hectares.[9] The Thai government, too, wishes to move closer to the heart of the boom by planting 1.6 million hectares by 2030, more than five times the hectarage of the early

Figure 4.1 Oil palm fruit, area harvested (1,000 ha), 1990–2008

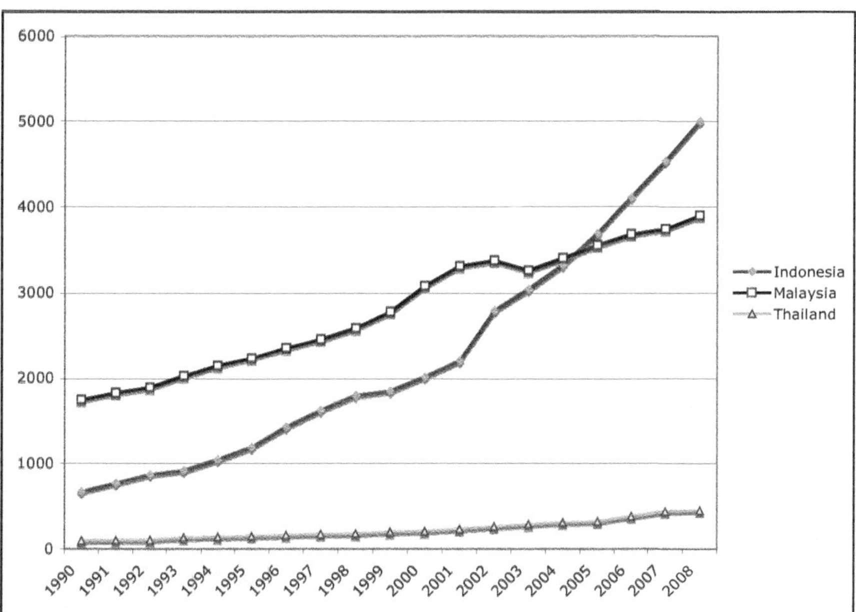

Source: FAOSTAT. Downloaded 3 Aug. 2010.

2000s.[10] Most dramatically, in May 2005 Indonesian President Yudhoyono announced a plan to create a 1.8 million hectare oil palm project along the border between Kalimantan and Malaysia, though this plan faced a storm of criticism (including from within the Indonesian government) and was quickly abandoned.[11]

To comprehend both the speed of the expansion that has taken place and these breathtaking plans for the future, we must begin with market demand. Palm oil, the most important product of the oil palm, is the world's second-most consumed edible oil after soybean oil and is also used in the production of other food products and manufactures (including soap).[12] Demand for palm oil for these uses is high and rising. Palm oil has also been tapped as a source of biofuel, the market for which could be almost limitless. From the demand side, then, there is little to stop the boom. It should also be noted that oil palm, unlike shrimp and coffee, has yet to experience a significant crash. Prices have been strong through most of the 2000s, though they did fall early in the century and briefly collapsed (along with the price of oil) in 2008.[13] Oil palm production has also not faced the kinds of disease problems that have plagued cocoa and shrimp farming.

While local conditions vary, two points can be made about the land on which oil palm is planted. First, oil palm generally replaces some level of forest cover. Niels Fold and Tina Svan Hansen argue that planting in Sabah and Sarawak has largely been on recently logged land,[14] and Hansen's study of land use change in the Niah River catchment area in Sarawak since the early 1970s substantiates this pattern of "successive waves of land cover changes where forest areas are logged, and afterwards gradually replaced by oil palm plantations".[15] In both Indonesia and Malaysia, too, companies have gained access to huge tracts of land by feigning interest in oil palm when they were really only looking for permission to cut the standing timber.[16] Second, oil palm land tends to be subject to ambiguous, contested and even contradictory tenure conditions. Much of it is subject to customary land claims with unclear or disputed status.[17] Recognizing that oil palm expansion often involves deforestation and runs roughshod over assertions of customary rights goes a long way towards explaining its intensely controversial nature.

Oil palm is farmed within the constraints of biology and ecology, and of the organizational frameworks that structure relations between land, labour, capital, technology and management. Oil palm trees reach maturity roughly three years after planting. Their fruit must be processed within 24–48 hours of harvest, and farms must thus be within relatively easy access of a crude palm oil mill. Economies of scale at these mills, in turn, require that they have access to planted areas of at least 4,000 hectares.[18] Within these constraints, production in Indonesia and Malaysia has gone on under a bewildering variety of organizational structures, with variations between states/provinces and between localities, and with major changes taking place over time.[19] Abstracting from the details, production takes place under the three usual frameworks: plantations/estates (private, state-owned or joint-venture), (more or less) independent smallholders (who may be supported by state agencies and/or corporations), and contract farming. On plantations, a company with access to land (generally through a state concession) hires labourers to carry out the work of planting, tending and harvesting. Most oil palm plantations in Southeast Asia are between 10,000 and 25,000 hectares in size.[20] The second model involves smallholders planting oil palm on their own land in hopes of selling the fruit to a nearby mill. Contract farming, most importantly under Indonesia's nucleus estate and smallholder schemes, has often been something of a mix of plantation and smallholder production, in that smallholder farms are combined with plantation-type areas on which the smallholders also work. Both plantations and contract farming projects may be state-owned or private, and contract farming may take place on land contributed by the smallholders or allocated through resettlement schemes. While diversity is an enduring element of oil palm production in Indonesia and Malaysia, both countries have seen strong neo-liberal trends away from direct state involvement in production and towards greater reliance on private companies since the 1980s.[21]

Oil palm is extremely popular with governments, and state actors in Indonesia and Malaysia have legitimated the crop in a striking range of ways. Stated reasons for creating smallholder schemes (including in oil palm) have included "the development of isolated rural areas, accumulation of property for landless farmers, better development and use of natural resources, creation of jobs, ensuring national food security, diversification of exports and stemming rural exodus", as well as the state's desire to settle shifting cultivators and to resettle people from densely populated cores.[22] Increasing state revenue and the desire to avoid political challenges from rural areas have also been policy goals of oil palm schemes. Perhaps we can underscore the importance of the tree by noting that it appears on the Indonesian Rp. 1,000 coin.[23] From the government perspective, there is little hesitation about the exclusions associated with this crop.

Sarawak

Malaysian oil palm production took place almost exclusively on the Peninsula until the late 1980s, when the crop began to take off on Sabah.[24] Sarawak joined the boom slightly later. In 1990, Sarawak accounted for 3 per cent and Sabah for 14 per cent of Malaysia's oil palm hectarage; by 2007, Sarawak had 15 per cent (or 664,612 hectares) to Sabah's 30 per cent.[25] In order to understand oil palm in Sarawak, we first need to know something about native customary land (NCL). Sarawak stands out from much of the rest of Southeast Asia in the extent of its legal acknowledgement of "native" land rights (though see the discussion of recent developments in the Philippines in Chapter 2), but that acknowledgement remains hedged about with restrictions. Formal state recognition of native rights to land and of some customary authority over land tenure and community territories began under the rule of the Brooke Rajahs, who took a "confusing and contradictory approach to land law and administration".[26] Brooke land law was consolidated into a single order in 1920, which classified land into "Town and Suburban Lands", "Country Lands" and "Native Holdings". A further consolidating order in 1931 distinguished for the first time between Native Areas and Mixed Zones.[27] After World War II Sarawak came under direct British rule, and in 1958 the colonial government passed the Land Code that still governs (with amendments) land relations in Sarawak. Four elements of the code are especially important for our purposes. First, it continued the racialization of land zoning. NCL is one of five land categories, and it is defined as land held not under title but under native customary rights; it can be held only by "natives", a category that includes Malays and Dayaks but excludes Chinese.[28] Second, NCL is not owned but held: it is licensed to natives but remains the property of the state. Third, to have NCL claims recognized, natives have to prove that they have occupied their land since before 1958;[29] in other words, the establishment of new NCL is meant to be frozen. Fourth, while NCL is restricted, it remains a legally valid concept that is recognized by the state (whether or not any particular claim is recognized).

Sarawak's government began seriously to promote oil palm in the 1970s and has taken various approaches to bringing together land, labour and capital. This history has been very helpfully untangled by R.A. Cramb.[30] In the decades before the boom, three public agencies were created to promote large-scale land development: the Sarawak Land Development Board (SLDB), a resettlement agency established in 1972 after the model of FELDA (see Chapter 2); the Sarawak Land Consolidation and Rehabilitation Authority (SALCRA), created in 1976 to undertake in situ development on NCL; and the Land Custody and Development Authority (LCDA), established in 1981 to initiate joint ventures between customary landholders and private developers in urban and rural areas. There was also a fourth mechanism for establishing large-scale production, which let private companies receive provisional leases to set up oil palm estates if they first resolved any land claims within the area.

Few companies took advantage of this possibility before the 1990s, and oil palm expansion under these schemes was limited.[31]

In the early 1990s, falling timber revenues and rising prices for palm oil prompted the state government to reinvigorate and transform this policy framework. In 1995, a new approach was introduced under the rubric of *Konsep Baru* (New Concept).[32] In explaining this policy, the government, and especially long-time Chief Minister Taib Mahmud, has consistently emphasized the concern that the Dayaks are being "left behind" in the race for development and that it is vital that their "idle" land be converted (for their own benefit) into a productive asset.[33] *Konsep Baru* takes a neo-liberal approach to this project. Oil palm expansion is now meant to take place through joint ventures (JVs) between the private sector, customary landholders, and either the SLDB or the LCDA. Landholders turn over their land to the government agency in trust, receiving some payment for it and a 30 per cent share in the JV that the agency forms with a private company. The agency consolidates the land (receiving 10 per cent of the venture), and a 60-year title to the land is issued to the JV. The company provides capital, expertise and management, and takes the remaining 60 per cent of the venture.[34] When the 60 years are up, the land is either sold on the holders' behalf or returned to the holders. Under the vision of *Konsep Baru*, then, both natives and the nation benefit from the unlocking of native land from the encumbrance of native customary rights, and the process, though it involves the state, is private and consensual.

With this institutional ground covered, we can move on to the central questions: What have been the implications of oil palm expansion for exclusion in Sarawak, and who have been the winners and losers in this process?[35] We begin with two qualifications of the neo-liberal vision sketched in the last paragraph. First, *Konsep Baru* was not the only change taking place in Sarawak's legal framework for land acquisition in the 1990s. Between 1994 and 2000, the Land Code was amended in ways that greatly facilitated the extinguishing of native customary rights to land.[36] Agricultural plantations, for instance, were classified as a public good, making it easier for the government to expropriate land to establish them.[37] By means of the power to regulate, then, the government made it easier for itself to take control of land to make it available for private oil palm development. Second, *Konsep Baru* schemes have not amounted to much. Cramb's careful breakdown of Sarawak's oil palm area as of October 2006 indicates that of the 581,040 hectares planted, only 5.7 per cent were *Konsep Baru* JVs. Nearly 75 per cent of the area, meanwhile, was controlled by private estates operating under provisional leases.[38] These estates, which are generally worked by labourers from Indonesia or farther afield, have mostly been established on land formerly zoned under the 1958 Land Code as Interior Area Land or Protected Forest Land (rather than NCL) but now reclassified as Mixed Zone Land. Native communities living on this land have generally not been recognized as possessing customary rights, either because they moved to the area after 1958 or because they cannot prove occupation since before that year.[39]

The spread of private estates is the real story in Sarawak's oil palm explosion. State and corporate actors have worked rapidly to expand plantations on native customary land as well as other land through a combination of legitimation, law and force. Cramb describes the arrival of such projects in the uplands from the years around 2000:

> longhouse communities who claimed customary rights to part or all of the land allocated for an oil palm estate often knew nothing of the granting of a provisional lease until bulldozers arrived to clear the area for planting. When they protested they were mostly ignored, given notice to leave the area or, in the worst cases, subjected to violence.[40]

How have upland people responded to these estates? Not surprisingly, the state/corporate land push has triggered "a widespread conflict over land".[41] Some villagers have resorted to direct physical opposition: blockading roads or interfering with bulldozers, for instance. Such clashes have, at times, escalated to violence. Natives have also made relatively effective use of the court system. Sarawak is unusual in this respect, both because the Malaysian judiciary is comparatively independent (in Southeast Asian terms) and because of the formal recognition given to NCL. Cramb notes, for instance, that as of 2000, 22 per cent of the oil palm estates in Sarawak were facing customary rights claims.[42] Cases against oil palm and timber plantation companies have resulted in favourable decisions for claimants. In the village of "Sungai Sulasah", for instance, a legal challenge "sufficed to push back the oil palm corporation". Court cases also, however, require financial resources and legal support (often from NGOs).[43] While a positive court decision may not effectively establish native customary rights against the weight of state and corporate power, the courts remain an important avenue of resistance in Sarawak in a way that is rare in Southeast Asia.[44]

People facing oil palm expansion have also tried to hold on to their land in less high-profile ways. One method has seen smallholders attempt to use the state's obsession with oil palm for their own benefit by planting the trees to protect their land against the state and corporations (and, indeed, their neighbours). Fadzilah Majid Cooke writes of her fieldwork site:

> villagers were vigorously planting their own oil palm, ahead of the companies that were known to be approaching the area. It was particularly important to

plant at the boundaries to neighbouring villages or close to roads, to make their claims visible to neighbours, government and companies. This kind of agriculture was something of a race against time because it was a way of staking a claim and making that claim visible—"ahead of the bulldozers".[45]

Jean-François Bissonnette reports that this strategy extends to villagers expanding the area that they have planted to other crops, a move that appeals to laws prohibiting disturbance of native territories under cultivation. He also notes a variation on the theme (and one that resonates with our discussion in Chapter 3) as villagers use the knowledge of state conservation priorities provided by NGOs to plant tree species that are protected from cutting.[46] Villagers have thus mobilized state priorities in their efforts to exclude others (including co-villagers) from their land.

The smallholder reaction to the oil palm boom has not been limited to resistance and rejection, however. Smallholders have participated in the boom through government agencies such as SALCRA, through JVs, and independently. There is significant unmet demand, for instance, for SALCRA's oil palm projects.[47] Dimbab Ngidang writes that his research group was surprised to discover that longhouse communities in the two villages they studied "seemed to be in a rush to participate" in nearby *Konsep Baru* JV pilot projects.[48] The motivation to join these schemes extends beyond a desire to grow oil palm—though that desire is often strong. In many areas, people join because they hope to gain access to project-related infrastructure provided by the state.[49] Smallholders and native groups also hope to consolidate their claims to their land by getting involved with oil palm, and have looked to the projects to provide training and managerial employment for their children.[50] Some also participate out of a fear of being labelled anti-government and facing violent reprisals.[51] Finally, Ngidang explains the desire to participate in part through surveys showing that only 17 per cent of respondents felt they understood how the schemes worked.[52]

Bissonnette's research in the village of "Sungai Binga" provides a compelling account of differing responses to a *Konsep Baru* JV. The prospect of the JV induced the government formally to delineate the village's native customary land—a prerequisite to *Konsep Baru* projects going ahead. However, as villagers became more familiar with a nearby *Konsep Baru* pilot project, and as they came to perceive the "power imbalances and the upper hand of the corporation", a substantial number rejected participation. Opponents of the scheme eventually broke away from the longhouse to form a new community. This group also engaged in confrontational tactics, including blockades and the expulsion of surveyors (leading to arrests), and filed a court case with the help of outside NGOs. While the legal tactics have met with some success, including an injunction to the Forestry Department to halt logging operations and a favourable court verdict in October 2006, other villagers have joined the project by signing individual contracts that transfer their land rights to the JV.[53]

While oil palm has been resisted in Sarawak, many smallholders there have independently planted oil palm on their own land. While this planting is partly a matter of claiming land, it is also motivated by the recognition that oil palm is highly profitable. In Sungai Sulasah, another site of resistance to plantation development, 8 of the 25 households Bissonnette surveyed were already relying exclusively on oil palm or planned to do so.[54] Similarly, when the Malaysian Palm Oil Board announced a project to support smallholders within reach of existing mills, it received close to 9,000 applications. State support for smallholder oil palm cultivation in fact goes back to the 1960s, but it has been minimal, and oil palm smallholdings have expanded in Sarawak largely without state support.[55] Cramb argues that the efforts of bodies such as SALCRA, the Department of Agriculture and MPOB to promote smallholder oil palm are actually being hindered by the state government, which, for all its talk of preventing the Dayaks from being "left behind", is much more supportive of *Konsep Baru* and private estates. While independent smallholders accounted for only 3.3 per cent of the planted oil palm area in Sarawak by October 2006, this had more to do with the politics of natural resources in Sarawak than with the intrinsic requirements of the crop.

The standard narrative of enclosure that we referred to in Chapter 1 fits very well with the story of oil palm expansion in Sarawak. State actors and private companies have worked together to seize enormous amounts of contested land for oil palm plantations worked largely by Indonesian migrants. Natives have found themselves dispossessed from land they claimed as their own and have resisted this dispossession. Plantation development has been lubricated by regulatory changes and appeals to development and (ironically) native *inclusion* into the mainstream of national life, but it is at times reducible to force. Even in this case, however, the straightforward "land-grab" narrative needs to be complicated. Natives are not necessarily "against oil palm". Many recognize it to be the most profitable land use on offer. Engagement with oil palm also offers benefits beyond those immediately associated with growing the crop: it facilitates the exclusion of other potential users (state agencies, corporations, neighbours) and can give access to infrastructure. And *Konsep Baru* schemes may look attractive in the context of deagrarianizing tendencies that make farming a less attractive livelihood and land use (see Chapter 5). As we have seen, there is potential for more bottom-up oil palm development in Sarawak when smallholders can engage with the crop on positive terms.

Prawndikes: Shrimp in Thailand

The sheer scale of oil palm expansion, in conjunction with its implications for displacement and deforestation, made it the most globally controversial boom crop in early 21st century Southeast Asia. During the 1990s, however, shrimp farming held the title. While shrimp has been farmed for centuries in parts of the region, the conjunction in the mid-1980s of technical breakthroughs, high demand, and support from states, corporations and international donors triggered an export-oriented shrimp rush (or "prawndike") that spread rapidly through coastal Southeast Asia.[56] As with oil palm, one reason the boom was so striking was the speed and the totality of landscape transformation. When shrimp farming arrived in new areas, huge stretches of coastline could be converted very rapidly. Mangrove forests, salt flats and agriculture gave way to shrimp ponds, almost as if the sea was invading the land. The boom's shock was intensified by the way shrimp farms could be abandoned just as quickly. Densely concentrated shrimp ponds are prone to devastating outbreaks of disease. In some areas shrimp farming collapsed just five years after it started, leaving behind salinated, abandoned land not good for much of anything. Even at the national level, the sector has been prone to booms and busts (see Figure 4.2), but national statistics hide even more dramatic crashes and relocation *within* countries. Shrimp farming also rapidly became notorious for a range of contentious and environmentally destructive dynamics, including land-grabbing, mangrove deforestation, pollution and violent conflict. To opponents, shrimp represented the triumph of Northern luxury consumption over livelihoods and sustainability in the South. Set against all this, however, was the fact that shrimp farming was not that hard to get into and was extremely lucrative—significantly more lucrative than either oil palm or coffee. The boom was driven not only by agriculturalists (who stood to make far more money raising shrimp than they could doing anything else with their land) but by professionals who gave up their jobs to dive into aquaculture.[57] One observer reminisced in the mid-1990s about the days when, in the area around Songkhla in southern Thailand, there "was gold on every wrist, new trucks zipping up and down Songkhla highway, and the karaoke bars were hopping".[58] The boom was important to states, too, as shrimp quickly became a major source of foreign exchange. In Thailand, the minister of agriculture called shrimp a "miracle animal" on these grounds.[59]

Even more than in the case of the other crops discussed in this chapter, our discussion of exclusion is complicated here by the remarkable diversity of Southeast Asian shrimp farming.[60] This diversity relates to the very wide range of levels of capital intensity and technological sophistication that has characterized the sector. Some shrimp farms are little more than lagoons

Figure 4.2 Brackishwater farmed shrimp production (tons), 1990–2008

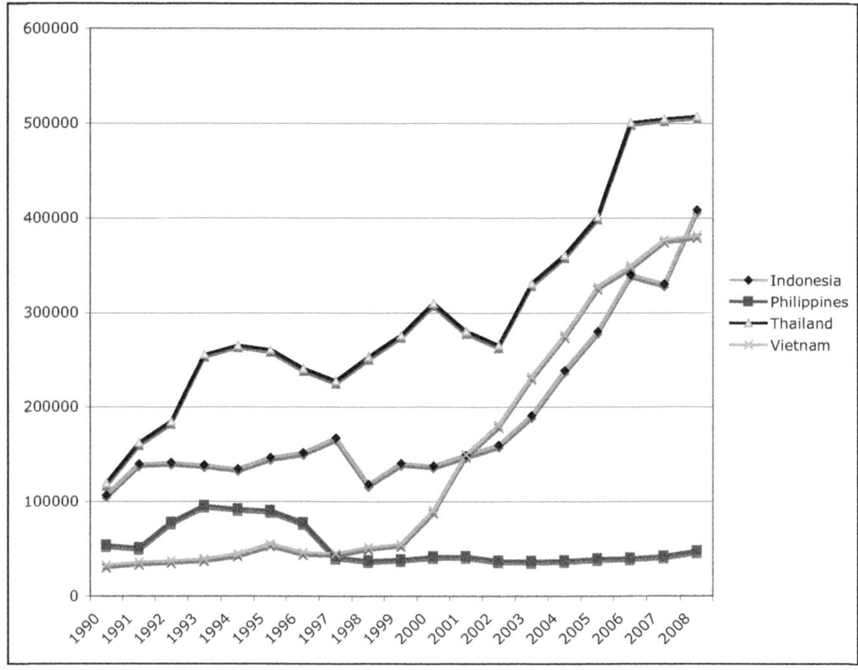

Source: FAO FIGIS. Downloaded 3 Aug. 2010.

with mud embankments built around them, with the shrimp and their food supplied by the tides; others use concrete tanks connected to sophisticated water circulation and oxygenation systems, high-tech imported feed, a range of chemicals, and broodstock that is certified disease-free and flown in from as far away as the United States. Some of this diversity is captured by the distinctions made between extensive, semi-intensive, intensive and super-intensive farms.[61] The fact that extensive and semi-intensive shrimp farming can be practised with limited capital and technical inputs has contributed to the establishment of large numbers of small, owner-operated farms, though even some of these farm intensively using sophisticated technology. There are also massive contract-growing schemes, with the biggest (in Indonesia) measuring tens of thousands of hectares.

The range of actors involved has been just as wide. Shrimp farms have been set up by fishers, by farmers, by professionals and by corporations. On the Philippine island of Negros, rich sugar planters rushed to establish them in the 1980s, while in Indonesia they have been created as part of state transmigration projects.[62] Shrimp farming has been promoted and supported, with subsidies and technology, by international and national aid agencies; by state extension agencies, fisheries departments and other bureaucracies; by academics and technicians; and by a range of corporate interests, from massive agribusinesses such as Charoen Pokphand (CP) in Thailand and San Miguel in the Philippines to small-scale traders and hatchery operators. It has been opposed, meanwhile, by NGOs (domestic and international), by academics and journalists, by the people negatively affected by it, and by certain state agencies. The diversity of technologies, organizational forms, and actors at work in Southeast Asian shrimp farming means that powers of exclusion have been exercised in a wide range of ways, and with quite different consequences, in this sector.

Booms and Busts in Thailand

With respect to Thailand, the country we focus on here, this complexity can be narrowed down a little—but only a little. While state and donor promotion of Thai shrimp farming began in the early 1970s, it was intensified around 1980,[63] and production took off circa 1985. The initial boom, which was located in the Upper Gulf area around Bangkok, quickly fell victim to diseases and to auto-pollution. By the early 1990s, the locus of the boom shifted to the eastern side of Thailand's southern peninsula, particularly the area around Surat Thani. The mid-1990s saw another crash (though production in the southeast recovered) and another round of relocation, this time to the west coast and inland. Thai shrimp farming has been characterized by small farms (averaging around 2 hectares) using intensive farming techniques, many of them owner-operated.[64] Compared to other major producers in Southeast Asia, Thai shrimp farms have seen more effective promotion by the state and agribusiness, but jurisdiction is fragmented and government agencies often work at cross purposes in their efforts to promote and regulate the sector.

All this diversity means that it is difficult to generalize about the mechanisms of exclusion at work. We thus focus on four different exclusionary processes that have shaped the trajectory of the Thai shrimp boom: the establishment of shrimp aquaculture in mangrove forests, the fight over shrimp in inland areas, the use of force to deter theft from ponds, and crashes. The first of these, the cutting of mangrove forests, is perhaps the most notorious exclusionary move in shrimp aquaculture.[65] In the early years of the Thai boom in particular, large areas of mangrove forest were cleared for shrimp. Mangrove deforestation creates a wide range of social, economic and ecological problems. Mangroves serve as a protective barrier against storms, tidal waves and erosion, as a nursery for many commercially valuable species of fish, and (often) as a source of livelihood for the people who live in and near them. All of this is lost when mangroves are cut down.[66] Mangrove conversion in Thailand fits comfortably into the classic narrative of enclosure and primitive accumulation. Because mangroves have not, historically, been economically valuable, the government was long content to allow small-scale fishing communities to live in and make use of them, though generally a de jure state claim to the area has gone along with this de facto lack of interest. With the shrimp boom, however, mangroves became potential sites for the production of an extremely valuable crop. The result was a frontier-style rush on the mangroves, as large areas were enclosed and converted by people (some local, some not) with the power to claim them, leaving former users excluded.

While there has been contention over these enclosures,[67] Thailand has not, for the most part, seen the violent confrontations over and organized resistance to shrimp that have marked the industry's expansion in countries such as Bangladesh, India and Honduras.[68] In many cases, "local" people have been as caught up in the shrimp boom as anyone else, and the involvement of locals has dampened conflict over mangrove clearance.[69] It is also important to take on board Andrew Walker's critiques of the "arborealization" of rural Thailand to note that the communities that live in or close to mangroves are not necessarily dependent on them for subsistence.[70] A series of detailed studies on mangroves and shrimp farming in different parts of Thailand indicated that opposition to aquaculture was strongest in the village where mangroves were most used by villagers, while in another village where this dependence did not exist, the mangroves were treated as open access.[71]

Mangrove conversion has been shaped by zoning, market forces and ideology. The zoning story is a familiar one. While worries over the rapid conversion of mangroves have prompted a number of efforts to tighten regulation and implement conservation measures,[72] state control remains tenuous. In part this is a consequence of the variety of state agencies that take often contradictory interests in the mangroves (though the Royal Forestry Department is meant to have exclusive jurisdiction, with the Department of Fisheries responsible for aquaculture), but it also derives from the area where zoning and market powers overlap: would-be shrimp farmers have often been able to bribe state officials to look the other way when they convert ostensibly off-limits mangroves.[73] This is not simply a story of corruption, incompetence or "lack of capacity". It also reflects the conflict between conservation as a state goal (see Chapter 3) and the powerful legitimating discourses promoted by state and corporate actors and international agencies that stress shrimp's role as a source of foreign exchange and development.

Shrimp farming's inland migration, the second phenomenon we discuss, reveals a different pattern. Until the mid-1990s, the need for access to salt water kept Thai shrimp farming located along the coasts. This changed when new culture techniques allowed shrimp to be grown at low levels of salinity. These "low salt" techniques, which rely on salt water trucked in from coastal areas, made shrimp farming an option for landholders in areas up to two hours' drive from the ocean.[74] After the 1997 collapse of the baht raised farm gate prices, shrimp farming exploded in some of Thailand's most fertile agricultural areas, most notably in the Chao Phraya Delta.[75] With rice agriculture decreasingly viable as a livelihood (see Chapter 5) and shrimp offering spectacular returns, many rice farmers with access to the capital required to build ponds and purchase infrastructure moved rapidly to join the boom.[76] As Brian Szuster and his co-authors write, "it is no wonder that rice farmers [were] tempted to gamble and try to earn in two or three years what would otherwise be a lifetime income."[77] Another study notes that some farmers unwilling or unable to farm shrimp themselves have leased their land to aspiring aquaculturalists for lucrative rents, and that corporations such as CP have again been crucial in pushing the boom inland.[78] However, inland shrimp farming has faced more effective (if less internationally high-profile) opposition than have shrimp farms in mangrove areas. As shrimp farms pumped salt water and effluent into local water supplies, rice farmers and other agriculturalists began mobilizing for a ban on inland shrimp aquaculture.[79] While mangrove forests had, prior to the shrimp boom, been seen by most state actors as useless waste, the prospect of environmental catastrophe unfolding in Thailand's rice basin allowed opponents of inland shrimp farms to invoke legitimating discourses of food security and of the central role of the rice-farming peasant in Thai life. Chao Phraya basin farmers, too, were politically better-connected than villagers living (often technically illegally) in the vicinity of mangrove forests. While relevant government agencies lined up on different sides of this issue, and inland shrimp farmers demonstrated in Bangkok, a ban on inland shrimp farming was imposed on 22 July 1998.[80] Governors were told to zone their province's land into brackishwater and freshwater areas, and to restrict shrimp farming to the latter. While the ban has not always been effective, this episode does illustrate the use of legitimating and regulatory power to prevent actors driven by market incentives (and the hope of a better life) from expanding boom crop production as much as they would have liked.[81]

A third exclusionary phenomenon in Thai shrimp farming—the attempt to prevent theft of this very high-value crop—stands out in the context of this book for its combination of intensity and short time horizon. Most of our analysis of exclusion deals with situations in which the people excluded from accessing land are trying to make some kind of long-term claim to it. Smallholders may want to prevent state and corporate actors from converting their land into an industrial park or a golf course, while state officials may try to keep people from farming in areas that have been zoned permanently as national parks. Powers of exclusion are also widely used, however, to try to prevent very short-term access to the land that does not necessarily call into question the shrimp farmer's claim to the land itself. The gated communities and other peri-urban developments that we discuss in Chapter 5, for instance, devote a good deal of effort to keeping out unauthorized personnel who might be thieves or vandals, or who might simply lower the tone of the area by looking too poor. In the case of boom crops, it is the value of the produce that makes theft a major concern. As Peter Vandergeest has pointed out, a pond full of shrimp is so valuable that a few hours with some nets and a pickup truck under the cover of darkness can yield thieves a harvest worth thousands of dollars. In response to this threat, shrimp farmers have invested in means of violence. Bigger farms often employ watchtowers, searchlights, armed guards and rifles; owner-operated farms tend not to use these capital-intensive techniques, but an owner might spend his nights on the farm with a gun when the shrimp get close to market size.[82]

Finally, there are the exclusionary effects of the dramatic collapses that have marked shrimp farming in Thailand and elsewhere. The main reason that shrimp farming can prosper in one part of a country while imploding in another is that crashes are caused not only by falling prices but by disease and pollution. Poorly designed water exchange systems, cheek-by-jowl siting of farms, and pollution not just from industrial, residential and agricultural sources but from other shrimp farms mean that epidemics can spread through a region's farms very quickly. The element of gambling common to all boom crops is particularly intense in shrimp farming, as both the potential pay-off and the risks are very high. In some cases, farmers simply hope that they can get in a couple of crops (less than a year's work) and cash in before their ponds succumb to disease. In addition, because the risk of disease increases with the number of farms in an area, the common tendency for early adopters of a boom crop to prosper while latecomers get burned is very much present in shrimp farming. People who have borrowed large amounts of money (from banks, from neighbours, from family) to farm shrimp have often been forced into bankruptcy when their crops failed.[83] The usual market-driven process of accumulation and dispossession, however, is given an extra twist by the damage that shrimp aquaculture does to the land itself. The salinization that results from growing shrimp in brackishwater ponds, for instance, can render land unsuitable for agriculture. While the land is not always permanently ruined, it can be very difficult to return land to agriculture after it has been used to farm shrimp, and as a result large areas are abandoned when shrimp booms crash.

Brewing Up Trouble: Coffee in Vietnam

During the 1990s, Vietnam went from being a negligible player in the global coffee trade to being the world's second-largest producer, after Brazil. The tale of this shift might almost have been invented as an illustration of a classic Southeast Asian agricultural commodity boom. In Vietnam's Central Highlands, an area with a complex and traumatic history of colonial intervention, wartime devastation and socialist (attempts at) reorganization, planting of Robusta coffee exploded around 1993 in response to a spike in the world coffee price and domestic institutional changes.[84] Between 1993 and 2000, the area of harvested coffee land in Vietnam increased more than 6.5 times (see Figure 4.3). This coffee was almost all for export, and the boom was driven almost exclusively by smallholders. Many of the early participants were ethnic Kinh migrants from the lowlands, and migration accelerated as coffee fever spread, but upland minority groups also took part. One report claims that during the boom, "almost all farmers" in Dak Lak province—the heart of the boom—were cultivating coffee, and Dang Dinh Trung's fieldwork in the Ede village of Buon Brieng A in Dak Lak (on which we will have frequent occasion to draw) shows similarly that by 1999 "almost every household had

Figure 4.3 Coffee, green, area harvested (1,000 ha), 1990–2008

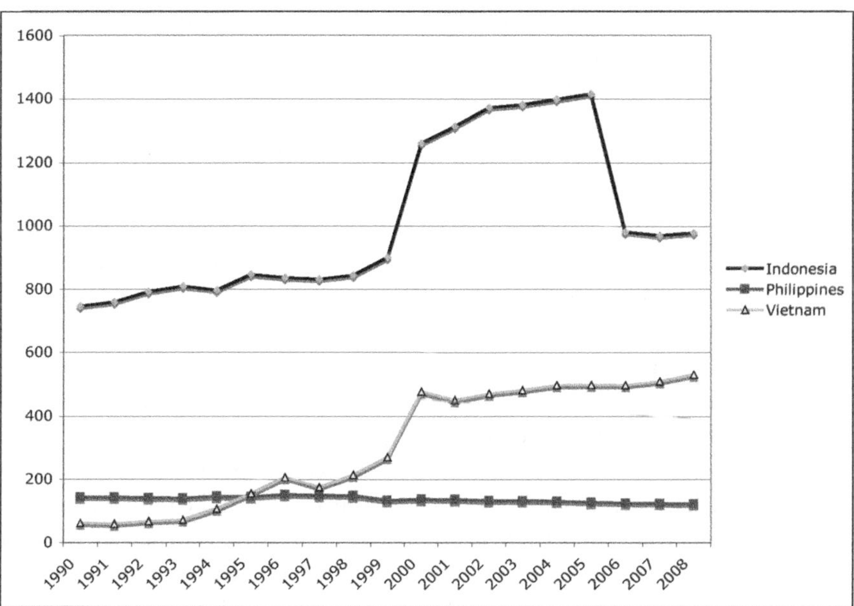

Source: FAOSTAT. Downloaded 3 Aug. 2010.

coffee fields".[85] The "dollar tree" and the "golden bean" made many people a lot of money, but the boom also profoundly changed life in the highlands. By the late 1990s, huge areas of garden, swidden and forestland had been converted to coffee, which dominated much of the landscape. As the boom made land much more valuable, the dynamics of access and exclusion also changed, becoming both more individualized and more contingent on the successful production of monocropped coffee. The extent to which highlanders had committed themselves to a single crop became painfully clear when the price of coffee suddenly collapsed in 1999—a collapse brought about in good measure, ironically, by Vietnam's very success in storming the world market. The shake-out that followed brought new forms of exclusion into relief. The market stands out amongst these, but, as we shall see, the other powers we highlight also continue to play a crucial role.

In order to understand coffee's impact on exclusion in the Central High-lands, it is helpful to begin with some information about the crop and the actors involved in growing it. The dynamics of both boom and bust varied, of course, by locality, but some generalizations can be made.[86] A striking element of Vietnamese coffee production (especially in comparison with Malaysian oil palm and Thai shrimp) is the limited farm-level involvement of the state and of large-scale agribusiness (though state-owned banks have extended substantial credit to the smallholder coffee sector).[87] While a 2004 World Bank Report emphasizes state provision of extension services, the effects are difficult to detect in the rest of the literature.[88] Almost none of the producers surveyed by the Information Centre for Agricultural and Rural Development and Oxfam in the early 2000s had received any extension services from the state.[89] Smallholders for the most part seem to have learned to grow coffee through a combination of copying their neighbours and trial and error.[90] These methods have clearly been successful: By the early years of the 21st century Vietnam was producing more coffee than Indonesia on around a third as much land.[91] Most smallholders sell their beans to traders.[92] State-owned plantations did continue to produce coffee during the 1990s, accounting for about 15 per cent of the country's crop, and in 1995 switched to contract relations with their growers.[93] Overall, however, Vietnam's coffee boom was driven more from below than the Malaysian oil palm rush or even the Thai "prawndike".

Certain characteristics of the crop help explain how this comparative independence has been possible. First, coffee processing can be undertaken very cheaply on the farm,[94] and there is no biological pressure (as there is with oil palm) to get the beans quickly to a central mill. Second, while coffee trees take four to five years to begin producing, farmers have been able to make it through the start-up period in part by intercropping with subsistence and cash crops (and, as we shall see, by taking out loans and selling assets).[95] However, while coffee has been a very smallholder-friendly crop in the Central Highlands, successful production requires access not only to land but to inputs, especially fertilizer and water. Several studies have found that access to inputs varies with ethnicity, with Kinh farmers better able to obtain loans (including from kin in the lowlands) and formal title, and recording higher yields per hectare.[96]

Where did the land for all this coffee come from, and how did people access it? In answering this question, we begin with the minority groups who were already in the Central Highlands before the boom, before considering the migrants who came during the 1970s, 1980s and 1990s. In Chapter 7 we outline the history of the peoples known collectively to the French as Montagnards; here, we can briefly point out that the coffee boom and bust represents the most recent of a series of dramatic and often traumatic changes in the highlands. The region suffered terribly during the wars, which saw the death of roughly a third of the population and the displacement of the great majority of villages.[97] After reunification, "the central government banned collective landownership, declared traditional tribal lands as state property available for redistribution and forbade nomadic slash-and-burn farming practices, forcing hill tribes to settle down".[98] Agricultural land was aggregated into cooperatives, and forest was assigned to State Forest Enterprises, while the main policy directed at upland minorities was the Fixed Cultivation and Sedentarization Program (1976–90).[99] Things changed again in 1981, when some aspects of control over land were reallocated from state farms to individual families, who nevertheless continued to have responsibilities to their cooperative. Further liberalization took place in 1988 with the passage of a law giving full control over field management to the household, a move that, one study argues, "opened the door towards private investments and went hand in hand with increased immigration and uncontrolled development of the coffee sector".[100] A new Land Law in 1993 then made it possible de jure for individuals to "buy, sell, inherit, mortgage and lease land use rights".[101] Smaller amounts of land zoned as forest have been reallocated from the state-owned enterprises, and in 1999 a pilot programme of forest devolution and reallocation was begun in Dak Lak province. This programme selectively granted group rights to villages, including the right to convert 5 per cent of the forest to cropland, while also imposing a duty to protect the forest.[102] None of these moves, of course, was completely effective, and actual land relations in the highlands often diverged significantly from the visions of government policy.

In the early 1990s, then, the tenure situation in the Central Highlands was in flux but had moved significantly towards household production. There was also substantial "available" land, in the form of swidden areas and forest, on which coffee could be planted. Dang Dinh Trung's research in an Ede village gives some sense of what happened in one location when coffee entered the picture. While the local state forest enterprise had encouraged villagers to plant coffee in the late 1980s, uptake was low.[103] However, as villagers became aware that pioneer planters (mostly, but not exclusively, migrants) were reaping windfall profits as a result of the price spike, a rush began to convert garden and swidden land to coffee. Dang Dinh Trung shows that while in 1994 the average household in Buon Brieng A had 3.4 hectares of land planted to swidden rice and 0.3 hectares to coffee, by 1999 the averages were 1.3 hectares for coffee and only 0.6 hectares for rice—indicating both a commitment to coffee that pushed household rice production far below the subsistence level and a sharp decrease in the amount of farmland per household, a point we shall return to below.[104] The consequences of the boom, which "made the Ede aware of the significance not only of coffee but also of land", were extensive and were magnified by the fact that growing coffee, a perennial that paid off in years, not months, required growers to assert clearer and more individualized claims to land. Households rushed to claim land that they were farming, land that they had farmed, and forestland. Longhouse extended family units rapidly splintered, in part because of the demands of the new crop, but also because the 1993 Land Law categorized longhouse groups as single "households" and subjected them to the same limit on landholding that nuclear families faced. One young woman explained these changes as follows:

> Formerly, we produced rice so we liked to work and share together. Now we plant coffee so we like to stay apart to earn a living. Living together is fun but we do not have enough land. In addition, some members are very lazy, while the others are industrious. Some know how to plant coffee while others do not. Some are clean and tidy, while the others are not. It's better for us to separate.[105]

While there was some tension in the village over who should get what land, Dang Dinh Trung argues that there was still sufficient land available for the issue to be resolved through negotiation, and that villagers' claims to land on which they had once planted rice but were now growing coffee were "finally accepted by all the village members". Only some of these claims, however, were covered by government-issued Red Book land titles.[106]

A second major change in Buon Brieng A and across the Central Highlands involved the sale or transfer of large amounts of land to migrants. Migrants—primarily, though not exclusively,[107] ethnic Kinh from the lowlands—had been pouring into the Central Highlands since long before the coffee boom. State officials, seeing the highlands as ripe for agricultural expansion, encouraged migration. Between reunification in 1975 and the *Doi*

Moi economic reforms of 1986, this movement took place primarily through a state-sponsored programme, but after 1986 unorganized, "spontaneous" migration dominated.[108] While these spontaneous movements peaked with the coffee boom (with 250,000 people officially moving to the Central Highlands between 1994 and 1999),[109] by the time the boom began large-scale migration into the Central Highlands had been taking place for more than 15 years. The region's population doubled to 3 million in the two decades from 1975. The change was even more dramatic in Dak Lak, where between 1975 and 1997 the population rose from 360,000 to 1.5 million while the share of minorities fell from 48 per cent to 20 per cent.[110]

Migrants gained access to land in various ways. Some received it from the state through the decollectivization, formalization and allocation programmes, including the allocation of plots to workers by state coffee farms.[111] Others cleared forest and set up as homesteaders. Stan Tan found in the late 1990s that the "social availability of land on the Dak Lak frontier is such that the migrants can 'instantly' convert it into abundant production of export coffee".[112] Yet other migrants have purchased land from people who were already in place, some of whom were selling in order to raise money for inputs so that they could grow coffee on their remaining land.[113] Land could be bought not only from minority people but also from earlier migrants; the overwhelming majority (over 95 per cent) of formal land transfers in Lam Ha District of Lam Dong province between 1994 and 2007 took place between Kinh.[114] Land also became available for rent.[115] Cramb *et al.*, finally, report an informal joint venture between a Bahnar family (which contributed 1.2 hectares of land) and a Kinh family (which contributed 1,000 coffee seedlings and some chemical fertilizer). This project ended badly when, after a dispute, the Kinh family beat up the head of the Bahnar household, sold off the land and refused to share the proceeds.[116]

Here again, a range of powers of exclusion was involved in the Central Highlands land rush. Given the persistence of swidden in the uplands, for instance, some of the "forest" being cleared by migrants was in fact fallow swidden land claimed by highland villagers. Migrants often either did not know of these claims or chose to ignore them. Dang Dinh Trung notes that the Ede came to feel insecure about land that they cultivated at some distance from their village, writing that "they often sold it at a cheap price because they were afraid that once it was left fallow, spontaneous migrants would occupy it".[117] Phan Trieu Giang has called attention to a related but distinct dynamic in Lam Ha in which Bahnar families would clear forest and plant it with coffee in order to sell it to migrants. This dynamic, ironically, derives largely from provincial and commune-level regulatory interventions that were meant to *protect* minorities from losing their land. Regulations that make it very difficult for minority people formally to transfer their land have encouraged them to sell informally to Kinh before moving on to clear more land at the frontier.[118] Transactions that might look straightforwardly like sales (and hence would seem to be driven by market forces) are in fact powerfully shaped by regulation and intimidation.

The boom did not last. For three years from late 1999, world coffee prices were so low that most Vietnamese coffee growers could not recoup even their variable costs of production.[119] While coffee production in Vietnam has not fallen along with prices, production and harvested area were more or less flat in the years after 2000 (see Figure 4.3). Smallholders reacted to the drop in prices in various ways, among them reducing the use of fertilizers and irrigation, burning or cutting down trees, planting new crops, borrowing money, seeking off-farm employment, and selling livestock.[120] Most farmers tried, however, to hang on to at least some of their coffee trees. One report states, "Almost all informants [...] claimed that, although coffee prices may continue to drop for some time, coffee has been, and is clearly, the most potentially profitable crop. Every coffee grower wanted to maintain their coffee farms and not replace them with other crops as advised."[121]

To what extent did the crisis in Vietnamese coffee production trigger a cycle of dispossession and accumulation amongst smallholders who could not cover their costs? One critical mechanism was debt. During the boom, many farmers took out loans from banks, private lenders, traders and relatives to invest in production and buy land. Servicing these debts became a major problem when the price of coffee plummeted.[122] As is so often the case in booms, the crash was especially devastating for farmers who got in late. Agergaard, Fold and Gough report that when the crash came to Dak Lak, "the more established farmers, who had enjoyed high prices during the mid 1990s, were considerably better cushioned by savings and less debt, compared with new entrants who had high loans and had gathered their initial harvest during low-price periods".[123] There is some dispute, or at least regional variation, regarding whether farmers responded to this crisis by selling land. While Dang Thanh Ha and Gerald Shively's 2003 research in Dak Lak found that none of their surveyed households had sold land,[124] NGO studies suggest that in at least some areas distress sales have taken place and more successful farmers have accumulated land.[125] One study found a classic pattern of dispossession and accumulation emerging from the crash:

> Those who managed to survive the crisis started accumulating cultivated land, while other unsuccessful growers had no choice but to lease out or sell off part of their land so they could continue to invest in the remains of their coffee farms. Once their own land had been sold, poor and indigenous people then sought out

other land deeper in the forest and in less favorable areas with steep slopes, low fertility, less water for irrigation and more difficult living conditions overall.[126]

Again, Dang Dinh Trung's research provides important insight into how the crisis played out at the village scale. He found that land sales had resulted from the subsistence gap faced by households that had difficulty meeting their consumption and investment needs because they had planted late or unsuccessfully, and thus "filled the financial gap by selling off means of production such as land and cattle".[127] His fieldwork, conducted during the crash, revealed further that a "striking contrast in landholdings between the poor and wealthy" had emerged in Buon Brieng A. While completely landless households had yet to appear, most poor households held only "a small area of rice land and a few coffee plants that had not borne fruit yet". Of the nine poorest households in the village, seven lacked land because they had sold it, primarily to Kinh. Indeed, a striking finding of Dang Dinh Trung's study is that while landholding in Buon Brieng A is becoming more stratified, richer Ede households are not accumulating land, as most sales are to outsiders. Trung also reports a finding relevant to our discussion of land titling in Chapter 2. Some households facing financial distress "tried to get a Red Book regardless of its cost" so that they could use this security to obtain bank loans.[128]

The implications of the crash for exclusion canvassed so far suggest a story in which markets (for coffee, for land, for credit) constitute the main powers of exclusion at work, with market dynamics shaped to some degree by the institutional and regulatory context of the Central Highlands (such as the partial and uneven extension of land use certificates).[129] However, the situation is more complicated than this. One other source of exclusion has been environmental. As people have brought more marginal land under coffee, the fertility of coffee land has begun to show signs of exhaustion, water supply has become a serious issue, and the ability of farmers to make a living on their land has been compromised. Deforestation, much of it caused by the coffee boom, has emerged as a major problem in the Central Highlands.[130] Even more strikingly, however, the collapse of coffee prices was a direct stimulus to a round of ethno-territorial conflict, most notably in 2001 and 2004, as minority groups attempted to mobilize powers of legitimation and force in order to reassert control over land. This conflict is covered in Chapter 7.

Conclusion

Crop booms that involve the conversion of large areas of land to a single land use are one of the key processes reconfiguring relations of land access and exclusion in rural Southeast Asia. Vast numbers of people have been caught up in these booms. People have converted their own land to the new crop and asserted stronger and often more individualized rights to it; migrated to acquire land and cultivate a boom crop far from home; and dealt with the fallout, which often includes loss of land through bankruptcy or environmental degradation. As our examination of oil palm, shrimp and coffee indicates, there are common features to crop booms, but also variation in the actors involved, the organizational frameworks used, key features of local history and geography, and the trajectories of boom and bust. Summarizing the powers deployed, we see both common patterns and significant diversity.

The market represents the most obvious power at work in a crop boom, given the role of export price fluctuations in shaping booms and busts. High prices attract would-be producers and investors to the crop, and they greatly increase the value of land, encouraging landholders to formalize their claims and initiating or intensifying the operation of land markets. Although it would be easy to fault in situ landholders for "self-exclusion" when they sell their land to corporations, incoming migrants or entrepreneurial farmers riding a boom on a path to accumulation, the decision to sell up is profoundly shaped by the interaction of the market with regulatory, forceful and legitimating power, a phenomenon we encounter again in Chapter 5.

States have made extensive use of regulatory power to zone and reallocate land for boom crop production. Regulatory changes regarding access to "native" and other land in Sarawak, exclusion from mangroves and agricultural land in Thailand, and the decollectivization and reallocation of agricultural land in Vietnam have been of the first order of importance in shaping the exclusionary dynamics of booms. States have also contributed to booms by selectively turning a blind eye to the infringement of regulation. Smallholders, too, positively invite intensified state regulation when they seek formal state recognition of land that booms have made more valuable, more scarce and potentially more contested, not least among kin and co-villagers who may previously have exercised less individualized claims. Relations of access to Thailand's mangrove forests, Sarawak's customary lands and Vietnam's highland swidden fields were radically changed by these crops booms, with the impetus coming, at least in part, from below.

Force comes out most clearly in our discussion of Sarawak, where communities have mobilized to defend their land from corporations and the state, but it has also been critical to the ways in which forests and swidden land were converted to coffee in Vietnam and to the defence of Thai shrimp ponds against theft. The use of market, regulatory and forceful powers, finally, is shaped by legitimating discourses that see boom crops as working to bring "remote" rural areas into the mainstream of national development and modernity, while also contributing to the national trade balance and the growth of GDP. While these discourses give boom crop producers a leg up in conflicts over land, they can also be appealed to by smallholders who have planted their land with oil palm or coffee as a way of preventing others from grabbing it. And while some state actors celebrate crop booms, others—especially those concerned with conservation—are much more nervous about them.

Our account also throws into sharp relief the ways in which smallholder boom crop producers feel exclusion's double edge. As we have shown, smallholders in Sarawak, Thailand and Vietnam have been, and remain, excited about growing these crops because of the real promise of prosperity that they hold out. But in order for them to participate, some practices, and some people, must be excluded. Sometimes the same people figure on both sides of the equation, as boom proceeds to bust for individual producers, or on a wide scale. For shrimp and coffee producers, joining the boom means undertaking substantial investment into inputs (usually requiring loans) and is demonstrably very risky. Price and disease crises can appear at any time to wipe out smallholders. Oil palm production has not seen crashes of the scale of those that have afflicted coffee or shrimp. However, smallholders in Sarawak face a different kind of risk: in order to grow oil palm, they need to enter into relations with state and corporate actors that are shaped by severe inequalities of power.

Notes

1. On the relationship between swidden and cash cropping more generally, see Cramb *et al.* 2009: 327–8, 337–9.
2. Hayami 2001.
3. Li 2010b: 390.
4. On the environmental problems associated with oil palm production, see Buckland 2005, McCarthy and Zen 2010.
5. *Economist* 2010.
6. Calculated from De Koninck, Bernard and Bissonnette 2011: Table 2.4.
7. McMorrow and Talip 2001.
8. Buckland 2005.
9. Wakker 2005.
10. Colchester *et al.* 2006: 21.
11. Wakker 2006; Potter 2009: 97–101.
12. Wakker 2005.
13. Potter 2008: 69; Potter 2010: 101–2.
14. Fold and Hansen 2007. See also Cramb 2011: 22.
15. Hansen 2005.
16. Colchester *et al.* 2006. See also Potter 2008: 81.

17. The complexities of customary and state claims to the political forest in Indonesia are outlined in Colchester *et al.* 2006; the situation in Sarawak is discussed below.
18. Wakker 2005.
19. For detailed accounts, see Cramb 2007a: Chapter 9; McCarthy and Cramb 2009; Sutton 2001; Colchester *et al.* 2006; Potter 2008.
20. Wakker 2005.
21. Acciaioli 2008: 94–6; McCarthy and Cramb 2009: 113–4; Sutton 2001: 91–2; Fold 2000.
22. Wakker 2005. On the state desire to use oil palm to diversify the economic base, see also Fold and Hansen 2007. On the narratives that legitimate oil palm, see also McCarthy and Cramb 2009.
23. On Sarawak, Fold and Hansen 2007.
24. Ibid.: 150.
25. De Koninck, Bernard and Bissonnette 2011: Table 2.4.
26. Cramb 2007a: 124.
27. Cramb 2007a: 129–30.
28. Cramb 2007b: 10.
29. Cooke 2002: 193.
30. Cramb 2007a: Chapter 9; Cramb 2007b.
31. Cramb 2007a: 268. By the late 1990s, 42,000 hectares of oil palm had been planted through SALCRA schemes (260–2). Cramb argues that a distinction can be drawn between SALCRA, which "proved acceptable to many Iban and Bidayuh communities in south-western Sarawak", and the LCDA and SLDB, which seem to have focused mainly on commercial gain (272).
32. See Cooke 2002, Ngidang 2002.
33. Cooke 2002: 189–90.
34. Cramb 2007a: 270.
35. Sarawak's experience on these points can be usefully compared with that of Kalimantan. See inter alia Acciaioli 2008; Potter 2008, 2009.
36. Cramb 2007b: 11–2.
37. Fold and Hansen 2007: 153.
38. Cramb 2007b: 23.
39. Fold and Hansen 2007: 154–5.
40. Cramb 2007a: 269.
41. Cramb 2007a: 269; Bissonnette 2011.
42. Cramb 2007a: 270.
43. Bissonnette 2011. See also Cramb 2007a: 270.
44. Up-to-date information on some of these court cases may be found at www.rengah.c2o.org.
45. Cooke 2002. See also Miyamoto 2006: 8.
46. Bissonnette 2011.
47. Cooke 2002: 194.
48. Ngidang 2002.
49. Fold and Hansen 2007, Ngidang 2002, Colchester *et al.* 2006.
50. On title, Ngidang 2002; Fold and Hansen 2007; Cooke 2002. On employment, Ngidang 2002.

51. Ngidang 2002.
52. Ibid.
53. Bissonnette 2011.
54. Ibid.
55. Cramb 2007b: 16–7.
56. On the causes of the boom, see Hall 2004b: 317–22. The term "prawndike" puns on the Klondike gold rush that took place in the Yukon in the 1890s.
57. Tokrisna 2004.
58. Simon Funge-Smith, quoted in Ahmed 1997.
59. *Bangkok Post*, 28 Aug. 1997, cited in Flaherty *et al.* 1999: 2054.
60. Hall 2004b.
61. Ibid.: 316.
62. Hall 2004b: 318, 328.
63. Flaherty *et al.* 1999: 2047.
64. See Tokrisna 2004.
65. Aksornkoae and Tokrisna 2004, Menasveta 1997. During the 1990s the government moved to protect mangroves from conversion.
66. Aksornkoae *et al.* 2004. After the Boxing Day tsunami of 2004, many scientists and activists argued that coastal areas where mangroves had been preserved were comparatively lightly hit. See Check 2005, Lewis 2005.
67. For a village-level study, see Johnson 2001.
68. Ahmed 1997, Stonich and Vandergeest 2001.
69. Hall 2003.
70. Walker 2004.
71. Aksornkoae, Sugunnasil and Sathirathai 2004.
72. Aksornkoae and Tokrisna 2004; Aksornkoae, Sugunnasil and Sathirathai 2004; Sugunnasil and Sathirathai 2004.
73. Sugunnasil and Sathirathai 2004.
74. Flaherty and Vandergeest 1998: 823.
75. Flaherty *et al.* 1999: 2049.
76. Flaherty *et al.* 1999: 2050; Szuster 2006: 90.
77. Szuster *et al.* 2003: 188.
78. Flaherty *et al.* 1999: 2055. For survey-based data on the backgrounds of inland shrimp farmers, see Szuster *et al.* 2003: 188–90.
79. There is some dispute over how much environmental damage inland farms in fact cause: see Mekhora and McCann 2003.
80. Flaherty *et al.* 1999: 2055.
81. Szuster 2006: 91. These rules thus try to preserve the state of affairs described in a song by The Soft Boys: "Where are the prawns? / Down by the sea."
82. Stonich and Vandergeest 2001. Lesley Potter has called attention to thefts from oil palm schemes in Kalimantan, though in this case the things being stolen were not fruit but rather fertilizer, herbicides, and even graders and bulldozers (Potter 2009: 118). See also our discussion of theft and arson in the Sulawesi cocoa boom in Chapter 6.
83. Sugunnasil and Sathirathai 2004. See also Flaherty *et al.* 1999: 2056.

84. The price went from around US$0.40/lb to US$1.40/lb in a few months in 1994 (Agergaard, Fold and Gough 2009: 136).
85. Asian Development Bank and ActionAid Vietnam 2003: 18; Trung 2003: 80. For other reports of the allure of coffee during the 1990s, see Ha and Shively 2004; D'haeze *et al.* 2005; Ha and Shively 2008: 317. For post-boom fieldwork showing between 81 per cent and 98 per cent of the households in four Dak Lak communes engaged in coffee production, see Agergaard, Fold and Gough 2009: 140.
86. See especially Agergaard, Fold and Gough 2009.
87. Giovannucci *et al.* 2004: xi.
88. Giovannucci *et al.* 2004: 3, 16–8. This report also points out (p. 16) that the Dak Lak Agricultural Extension Center had 20 coffee specialists providing services to an area with half a million coffee farmers.
89. Information Centre for Agricultural and Rural Development and Oxfam 2002: 19, 30.
90. Tan 2000.
91. FAO statistics.
92. ICARD and Oxfam 2002: 15. On the role of traders, see also Agergaard, Fold and Gough 2009: 135, 140–3.
93. Giovannucci *et al.* 2004: 11.
94. Tan 2000.
95. D'haeze *et al.* 2005.
96. Doutriaux, Geisler and Shively 2008; Giang 2010; Agergaard, Fold and Gough 2009: 143; Ha and Shively 2008: 322–3.
97. Hickey 1988, cited in Trung 2003: 75.
98. D'haeze *et al.* 2005.
99. Thanh and Sikor 2006; Trung 2003: 76.
100. D'haeze *et al.* 2005.
101. Doutriaux, Geisler and Shively 2008: 535–6.
102. Thanh and Sikor 2006, Sikor and Thanh 2007.
103. Trung 2003: 79–80.
104. Trung 2003: 80–1.
105. Trung 2003: 84–5.
106. Trung 2003: 80, 83.
107. De Koninck 2006: 50.
108. For a critique of dichotomous understandings of "planned" and "spontaneous" migration in Vietnam, see Zhang *et al.* 2006.
109. D'haeze *et al.* 2005; Winkels 2008: 34.
110. Hardy 2000; D'haeze *et al.* 2005.
111. Doutriaux, Geisler and Shively 2008: 535.
112. Tan 2000. On migrant coffee production on forest-zoned land that the State Forest Enterprises ostensibly held but were in practice not able to control, see ADB and ActionAid Vietnam 2003: 53.
113. Cramb *et al.* 2009: 340.
114. Giang 2010: 6.
115. ADB and ActionAid Vietnam 2003: 50; ICARD and Oxfam 2002: 32.

116. Cramb *et al.* 2009: 340.
117. Trung 2003: 93.
118. Giang 2010: 4–8. Similar dynamics have been identified in cocoa and coffee farming in Sulawesi: see Ruf and Yoddang 2001: 132; Ruf, Yoddang and Ardhy 2001: 183.
119. Ha and Shively 2008: 314.
120. Ha and Shively 2008: 318–9. On destroying trees, see also Greenfield 2002; ICARD and Oxfam 2002: 37–8.
121. ADB and ActionAid Vietnam 2003: 21.
122. Eakin, Winkels and Sendzimir 2009: 404; Winkels 2008: 35.
123. Agergaard, Fold and Gough 2009: 143.
124. Ha and Shively 2008: 319.
125. ADB and ActionAid Vietnam 2003: 14, 19, 21; ICARD and Oxfam 2002: 33.
126. ADB and ActionAid Vietnam 2003: 14.
127. Trung 2003: 93.
128. Trung 2003: 90–4.
129. See Cramb *et al.* 2004: 262.
130. See D'haeze *et al.* 2005, Winkels 2008. On coffee farming as a driver of deforestation in Sumatra, see Gaveau *et al.* 2009.

5

Post-agrarian Exclusions:
Land Conversion

Rapid economic growth in Southeast Asia has created strong demand for land for industry, housing, commerce, infrastructure, tourism and speculation, and huge amounts of land have been converted from agricultural to non-agricultural uses. In this chapter, we ask how the people who once farmed this land have shared in or been excluded from the prosperity new forms of growth have generated. There is extensive evidence from across Southeast Asia that farmers would like to get out of agriculture themselves and, even more, that they hope their children will not become farmers. Land conversion seems to offer an attractive opportunity to make this move. Yet this opportunity is fraught with risks, presenting dilemmas for smallholders and for other actors concerned about their welfare.

As with crop booms, the central trigger for the sale of land and its subsequent conversion has been rising prices—in this case, of land in the vicinity of urban areas, industrial zones and tourist hot spots. Many farmers have been keen to take advantage of rising land values by selling some or all of their land, and some have clearly benefited from doing so. These sales, however, take place in land markets that bear little resemblance to the orderly, transparent transactions envisaged by the World Bank (see Chapter 2). The risk for smallholders stems from the way that rising land values, in some respects so beneficial to smallholders, also prompt other actors to use the powers of exclusion at their disposal to lay claim to land. State officials rezone agricultural land to commerce or industry in a bid to raise local tax revenues, or simply expropriate it. Powerful actors use coercion to grab valuable land. And agriculture is given low priority in the discourses of development and modernity that frame debates over land conversion. Smallholders thus must struggle to work out the place of land in their lives under conditions changing at bewildering speed, as the very processes that hold out the promise of windfall gains and a transition out of agriculture also threaten farmers' claims to their land.

The question of land conversion and, even more, a consideration of the forces behind it need to be understood in the context of *deagrarianization*. Rapid industrialization and economic growth transformed Southeast Asian economies and societies in the second half of the 20th century, and agriculture's contribution to national GDPs has dropped dramatically relative to industry and services (see Table 1.1). As Jonathan Rigg shows in his 2001 book *More than the Soil*, moreover, this diversification away from farming has taken place not only at the national scale but in rural areas themselves. Deagrarianization, for Rigg, is made up of a variety of strands: a decreasing role for agriculture in rural people's occupations and income sources; "social reidentification", as people come to think of themselves less as farmers, and imagine futures outside agriculture; increased movement between rural and urban areas; and "spatial interpenetration", as urban and rural land uses become increasingly intermixed.[1] Deagrarianization is most striking when viewed in terms of occupation. The statistics in Table 1.1 that we referenced above show that the percentage of Southeast Asians classified as gaining their livelihoods from agriculture has been falling rapidly (though, as we also noted, the absolute number of people so classified rose between 1980 and 2005 in most countries). Although there is significant variation across the region, Rigg shows that villagers in many parts of Southeast Asia have a low, and declining, dependence on farming. As rural youth turn to industrial and service sector work in their home countries and abroad, farming is becoming the province of the middle-aged and elderly.[2] Deagrarianization may be *least* visible, on the other hand, in terms of land use. In the province of Nonthaburi, just north of Bangkok, for instance, agriculture in 1995 accounted for 71 per cent of land use while making up a mere 4.2 per cent of Gross Provincial Product.[3]

In this chapter, we examine land conversion (accomplished, desired and foiled) in Southeast Asia in the context of both deagrarianizing tendencies and the continued importance of agricultural land and livelihoods. The first section of the chapter is devoted to *peri-urbanization*, the process by which massive belts of mixed "urban" and "rural" land use have been created in the extended metropolitan regions (EMRs) of Southeast Asian cities. We lay out the powers at work in processes of peri-urbanization in general terms, then explore how they play out in a particular local and national configuration: Cavite, a peri-urban province near Manila. Building on this study, we offer a shorter examination of land conversion for tourism, focusing on the responses of villagers in Bali to the growth of tourism on the island, and on the struggles over land use that have accompanied the expansion of tourism around Angkor in Cambodia. The chapter concludes with a discussion of infrastructure, particularly dams, with special attention to the Hoa Binh Dam in northern Vietnam. This final section provides a contrast to the rest of the chapter, in that from the perspective of smallholders, dams do not present the dilemma of opportunity versus risk: rather, most dam projects are straightforwardly, and often catastrophically, negative for affected people and present them with no choice.

Peri-urbanization

Peri-urbanization is one of the most striking and powerful processes by which agricultural land is being converted to non-agricultural uses in Southeast Asia. Peri-urban or *desakota*[4] areas have been described in various ways,[5] but for our purposes their most important characteristic is the intermingling of what are usually taken to be urban land uses (industry, housing, commerce and so on) with agricultural production. While peri-urban zones also surround "second-tier" Southeast Asian cities such as Chiang Mai and Surabaya, most research focuses on the biggest cities, and we will largely limit our discussion to five EMRs: those around Bangkok (including the Eastern Seaboard zone), Ho Chi Minh City and Hanoi; the Jabotabek region surrounding Jakarta; and the Calabarzon zone centred on Manila, which incorporates Cavite.[6] These peri-urban regions are vast, stretching across areas measured in hundreds or even thousands of square kilometres and encompassing populations that, for Manila, Jakarta and Bangkok, top 15 million people.[7] *Desakota* zones are also visually striking and indeed incongruous. Photographs of rice paddies in the shadow of warehouses, or of water buffaloes grazing in front of high-rise apartment buildings, illustrate the intermingling of divergent land uses that defines peri-urbanization. Despite the often dramatic understatement of land conversion in official statistics,[8] it is clear that a great deal of this urban expansion has taken place on what was once agricultural land.

While peri-urbanization is complex and locally specific, we can identify four unifying processes that justify speaking of it as a regional phenomenon.[9] First, population and employment in Southeast Asia's cities are growing much more quickly in the peripheries than they are at the centre. The peri-urban province of Nonthaburi, for instance, had the highest population growth of any Thai province in the early 1990s.[10] Second, peri-urbanization has been driven by global and regional capital. The export-oriented foreign direct investment (FDI) in manufacturing that Southeast Asia has received since the mid-1980s has gone primarily to peri-urban areas, which combine infrastructure, low land prices, and relative proximity to the services available in central business districts. The economic stimulus generated by FDI and domestic investment has transformed peri-urban areas into "the locus of job creation" for Southeast Asia.[11] External private capital also helped to fuel the pre-1997 real estate boom in Indonesia, Malaysia, the Philippines and Thailand, and those countries were all hit hard when the boom collapsed in 1997, leaving behind abandoned high-end residential complexes in the midst of paddy fields as an enduring symbol of Southeast Asia's burst bubble.

Third, government policy has supported peri-urbanization. National governments have created master plans that attempt to guide and facilitate peri-urban development, for instance through the plans for the Eastern Seaboard in Thailand, the multimedia corridor south of Kuala Lumpur, and the Master Plan for Jabotabek that delineated desired primary and secondary growth centres.[12] States provide transportation, communications, water and other infrastructure, often with support from (especially Japanese) aid. They also provide land, whether public or expropriated.[13] They establish incentives for industrial development that encourage industry to locate outside core urban areas.[14] National governments share with other levels of government the powers to zone land and approve conversion, and the administrative decentralization undertaken across much of Southeast Asia during the 1990s and 2000s has relocated much regulatory power.[15] Fourth, and finally, *desakota* areas have seen widespread contention over land. Rising inequality, and often outright displacement, have gone with the conversion of land to new uses. Conflict is linked to the fact that peri-urbanization has so often been an unregulated and indeed illegal process, one described by the Indonesian academic Tommy Firman as "out of control".[16]

In analyzing the powers at work in the process of peri-urbanization, it is appropriate to begin with the market. As demand for industrial, commercial and residential land has increased, reported land prices in the areas around major cities have skyrocketed. In Phimonrat in Nonthaburi, reported average land prices rose from roughly 30,000 baht per *rai* in 1985 to 70,000–100,000 baht in 1990, before hitting 2 million–3 million baht in 1995.[17] The prospect of receiving prices like these creates an enormous incentive for landowners to sell up, and for tenants to exchange their tenancy rights for compensation. With smallholder agriculture facing severe problems, it is not surprising that many studies of peri-urbanization show that farmers and tenants are keen to sell.[18] One author even provides accounts from Indonesia of farmers destroying irrigation works on their land in order to justify conversion to authorities.[19] Rising prices have also, however, prompted farmers to hang on to some of their land as a store of wealth, in hopes that they can cash in when prices are higher or on retirement, or hand the land down to their children.

Mark Askew's mid-1990s research in the Thai sub-districts (*tambon*) of Bang Khanun and Phimonrat in Nonthaburi, and To Xuan Phuc's 2003–4 fieldwork in the upland Dao village of Hop Son in Vietnam provide valuable studies of these smallholder decisions to sell or hold land. In Nonthaburi, the arrival of manufacturing investment (from the mid-1980s) and residential development (from the mid-1990s) sparked the boom in land prices noted above.[20] Askew's work focuses on land sales as one component of broader livelihood and status-seeking strategies practised by households against the background not only of a real estate boom but of decades of occupational deagrarianization. His research among the market gardeners of Bang Khanun and the rice farmers and fruit tree growers of Phimonrat shows that households have been quick to sell off (some of) their land in response to market incentives. This alacrity, along with statistics showing the falling importance of agriculture as a component of household income in the two *tambon*, might suggest that agricultural land is no longer of much importance in these areas, but Askew argues that this is not the case. This is in part because the receipt of land through inheritance "has played the preeminent role in determining life chances", but also because the possession of land remains critically important to household status.[21] Askew emphasizes the careful use of land sales as part of a strategy of diversifying income sources and accessing capital for purposes such as homebuilding, debt repayment and education. He also notes the "generally shared view that family landholdings must be preserved wisely for the twin purposes of future sale for capital accumulation and sustainable income generation for children in the future".[22]

While Askew provides important insights on the long-term background to land conversion, To Xuan Phuc's research highlights the extraordinary speed with which land markets can transform a community.[23] Hop Son, about 60 kilometres from Hanoi, had been little touched by the land market until late 2003, when people from Hanoi began buying up land to construct weekend homes or to hold for speculative purposes. A year later, virtually all of the village land easily accessible by road had been sold. Many farmers sold not only their farmland but the plots on which their own houses stood. Villagers may be divided into several groups based on their response to this opportunity. Some (mostly younger) households spent most of their windfall on consumption goods and were already falling into debt and relying on wage labour within a year or two of sale. Others, older and wealthier, invested part of their proceeds in production goods, such as paddy land and livestock. Another group, the wealthiest, refused to sell up, and claimed that those who had done so were setting themselves up for trouble. Finally, a "marginalized group" was "totally excluded from the market" for various reasons, including holding plots in inconvenient areas, lacking title to their land, and having plots too small to sell. City people usually would not buy land with unclear legal status, a finding relevant to our discussion in Chapter 2. In the space of a year, these decisions created in Hop Son what To Xuan Phuc calls a "hybrid landscape" mixing productive and leisured pursuits.

The preceding paragraphs call to mind a vision of land being allocated through the market to its most efficient and profitable use, a process that may have negative repercussions for some groups (for instance, landless labourers losing agricultural work),[24] and that may be driven by debt and distress on the seller's side, but which takes place through mutually agreed transactions. Such a narrative suggests that people are excluding themselves from land by choice. This account needs to be complicated, and much of the rest of this section will be devoted to complicating it. One factor that highlights the power imbalances involved in these transactions, but which does not necessarily involve the mobilization of other powers of exclusion, is the role of land brokers. In peri-urbanizing areas characterized by large numbers of small plots, assembling land into large enough parcels for developers' purposes is time-consuming and costly.[25] Brokers have facilitated this process in part by using superior access to information to purchase land from smallholders at below-market prices. A report on the Jakarta area in the late 1980s, for instance, noted that "[u]ninformed small farmers on the urban fringe and homeowners in informal settlements [...] on centrally located land may sell their land at a fraction of its market price. Speculators who have bought up land in anticipation of development may hold out for prices many times higher."[26] At Hop Son, too, a land broker provided To Xuan Phuc with a vivid account of how he and another broker had colluded to drive down the purchase price of a plot of land.[27]

The need to consolidate parcels has been a key factor encouraging various actors to turn to non-market powers of exclusion. One of these is the power to regulate. The regulations that shape land markets often tilt the playing field on which the market game takes place. A series of studies from the 1990s on the rules governing land development in the Jabotabek region provide a striking example. Under this system, the government awards location permits giving a single developer the right to buy and develop land within a certain area. In some cases, these permits cover thousands of hectares. This system helps developers solve the problem of assembling extremely fragmented landholdings into larger parcels, and indeed Michael Leaf argues that it was "created specifically to feed urban lands into the corporate development sector".[28] However, the monopsony power this system gives developers has created enormous problems for local smallholders and, arguably, for Jakarta's urban development. One of the most striking consequences of the location permits is that huge quantities of land have been kept off the market. Firman writes that just before the 1997 crisis, developers controlled about 70 per cent of the undeveloped land around Jakarta but had developed only 12 per cent of it.[29] This situation resulted from an incentive structure that encouraged developers to seek permits for as much land as they could (whether with plans to build on it or for speculative purposes) but that also, by eliminating competition, allowed them to take their time over actually buying it. Bruce Ferguson and Michael Hoffman claim that the main purchasing strategy of one Jakarta-area developer was to wait for "landowners with pressing financial needs such as weddings or religious pilgrimages to offer their land at lower prices".[30] The system thus essentially excludes smallholders from the market for land, depresses the price they receive when the company does offer to buy, and encourages distress sales. As one might expect, this approach has generated plenty of conflict.

The Philippines' Comprehensive Agrarian Reform Program (CARP; see Chapter 2) has created a different, but perhaps equally perverse, conversion dynamic. One might expect the 1988 CARP law to protect agricultural land, given that it forbids the conversion of rice and corn lands deemed eligible for redistribution and gives regulatory authority over the conversion of all agricultural land to the Department of Agrarian Reform. However, as Philip Kelly has argued, the existence of CARP has often *accelerated* conversion. Landlords have rushed to get rid of their tenants in order to avoid the risk of having their land subjected to agrarian reform. They have been helped in this by a series of national-level decisions facilitating the removal of certain kinds of land from CARP jurisdiction.[31] Municipal governments have often assisted in the process, encouraged by the political power of landlords and developers, by the "market power" of bribes and kickbacks, and by their strong vested interest in seeing conversion occur so that they can collect fees and increase the tax base.[32] A striking consequence of CARP is thus that landlords are encouraged to remove tenants from their land even when they have no immediate intentions to sell or convert it. As in Jabotabek, peri-urbanization in Calabarzon has been marked by the peculiar phenomenon of large quantities of land being made idle at exactly the time that land is rapidly increasing in price (see Chapter 2).[33]

A second area where the use of regulatory power comes into play is in the zoning and rezoning of land. Peri-urban development generally requires, at least de jure, the reclassification to industrial, commercial or residential use of land previously zoned for agriculture, forestry or conservation. Indeed, these latter categories are designed to exclude developers and industry from the land in question and thus represent a barrier that they need to overcome. In the Philippines, developers have been able to remove these obstacles through direct control of city governments, a method encouraged by the country's combination of vigorous local electoral politics and decentralized administration.[34] We will encounter this phenomenon in our study of Cavite, but Edsel Sajor's account of the developer-dominated city council in Cebu provides another example of local governments helping private developers by rewriting land regulations, rezoning agricultural and protected land for development, and applying "spot zoning" to extract parcels of land from the wider zoned area of which they had been a part.[35]

The more common approach to dealing with zoning obstacles in peri-urban Southeast Asia, however, is to evade them. Leaf cites estimates that 70–90 per cent of all urban construction in Hanoi in the early 1990s to mid-1990s was carried out "without formal permits, or in most cases, even without proper authorization of land use rights" and was "perforce illegal"—though also "not entirely unauthorized, as typically a household will make arrangements with local ward officials in order to undertake redevelopment or market exchange of their properties".[36] More broadly, peri-urban areas are marked by ineffective or essentially absent planning and regulation, with zoning and land use maps often rudimentary or non-existent,[37] extensive development on land technically out of bounds for conversion,[38] and planning departments that are underfunded and lack political independence.[39] A large Jakarta-based developer commented to Ferguson and Hoffman that "[z]oning has little connection with actual land use", a comment that can be generalized to the region.[40] While the power to exclude through regulation is certainly important, then, its importance should not be exaggerated.

Land conversion in peri-urban Southeast Asia is also marked by violence and intimidation. The actors using force include military and police (working on behalf of the state and/or for private purposes), private gangs, militias, criminal organizations, and the people facing dispossession (and, in the Philippines, the NPA, an armed Communist insurgent group). Cases of intimidation pepper the literature. Firman reports from Jabotabek on "several protests and a struggle among land owners for fairer compensation" that generally ended with smallholders finding themselves "powerless against developers who often directly and indirectly intimidate them". A study carried out in areas undergoing conversion on the fringe of Bandung showed two-thirds of households stating that they had given up their land, even with poor compensation, because they felt that they had no choice.[41] One strategy used by developers was to build walls around the plots of people who wouldn't sell out.[42] In Thailand, the role of land broker is often played by local "godfathers", or *chao pho*, who hold enormous social and political power and are known to have access to substantial means of violence.[43] And as our discussion of Cavite will illustrate, there has been no shortage of violence in peri-urban Manila. H.W. Dick and P.J. Rimmer contend, finally, that a key force driving the development of gated suburbs in Southeast Asia is fear: the fear felt by middle-income and elite strata as a result of the highly skewed income distributions of Asian cities and the inability of the state to maintain public security.[44] The sense of insecurity among the elite is compounded for ethnic Chinese minorities, who have been the target of racialized attacks.

Finally, legitimation, especially the meaning and moral valence ascribed to land, is at the heart of land conversion. For state officials, while the conversion of agricultural land raises concerns about national food security, the overwhelming preference is for non-agricultural land uses, whether industrial, tourist, commercial or residential. This preference is expressed, across the region, in familiar languages of development, modernity and national benefit. It is important to note that these discourses do, under certain conditions, have some resonance with smallholders. Benedict Kerkvliet reports of villagers living close to Hanoi that they were prepared to relinquish some land for the construction of a highway "because they could see that [the] project benefited the nation", even though they "vigorously objected when provincial and district authorities compelled them to give up their fields so that several joint venture companies could establish an industrial zone".[45] New town development, too, has been driven by more than the search for profit. In an article subtitled "A Concurrence of Economics and Ideology", Leaf argues that the suburbanization of Jakarta is fundamentally a result of government policy "rather than merely being the natural result of market forces". At the core of the government's efforts has been the idea that Jakarta, as the nation's capital, "must necessarily express the power and centrality of the state", and that to fulfill this role it must be able to hold its head up as a world-class city "on a par with the Dallases, the Las Vegases and the Los Angeleses which fill the nation's television screens".[46]

The moral claims and understandings articulated by the smallholders and tenants who face the conversion of their land are discussed further in Chapter 7. Here, we take up the lived experience of deagrarianization and the ways it shapes how farmers see their relationship to the land. Two crucial aspects of this are the sense that urbanization is inevitable and the hope, frequently expressed by farmers, that their children will escape from farming.[47] To give just one example, Rigg and Albert Salamanca report that in interviews with aquaculturalists in peri-urban Bangkok, Phnom Penh, Ho Chi Minh City and Hanoi, every respondent expressed a preference that their children not follow in their footsteps.[48] The higher cultural capital (and, usually, income) associated with non-agricultural work, and the sense that agriculture is receding into the past, form a crucial context for understanding conversion decisions.

However, Askew's ethnography of Nonthaburi shows that deagrarianization is not the whole picture, or rather that it must be understood in its complexity. Peri-urbanization is experienced as double-edged when people regret the transformation of rural life while also participating in it and benefiting from the opportunities it brings. Askew provides a compelling example:

> As one gardener of Bang Kruai District, Nonthaburi, remarked to me: 'sang-khom muang khao ma kin mot loei' (urban society has come in and consumed everything). Yet this is only part of the story. Together with his neighbours, this same gardener attended a public hearing on the new master plan for Nonthaburi (in 1996) and vehemently opposed the classification of his land as 'rural and agricultural' on the grounds that the land values of his holdings would be depressed. The same gardener vowed that his children would never work in agriculture [...].[49]

In concluding this discussion of the powers involved in land conversion, we emphasize again that these powers are often fused in practice. Two examples forcefully make this point. The first, which Firman has explored in Indonesia and Gavin Shatkin in the Philippines, involves the extent to which it may not be just new town development that is being carried out by corporations, but the entire process of urban planning, something usually understood to be the role of government.[50] A second example is the way in which, in the Philippines and Thailand especially, the rapid expansion of peri-urban industrial and residential development has been inseparable from the political and economic power of local strongmen or "bosses" who facilitate land acquisition, finesse regulatory issues and cow opposition.[51] Shatkin has traced the way in which local Thai strongmen have been able to take advantage of industrialization, urbanization, democratization and decentralization to amass unprecedented power at the local and indeed national scales. Focusing on the career of the notorious Somchai Khunpleum (aka Kamnan Pho) of Chonburi, Shatkin analyzes his ability to use "economic influence and raw power to consolidate land, foster consensus around a development agenda, and suppress political opposition to development". Shatkin argues that the *chao pho* have played an underappreciated role in the emergence of export-oriented industrialization in Thailand.[52]

Cavite

This discussion of bosses provides the perfect bridge to Cavite. While we have focused up to this point on regional commonalities, the relationship between peri-urbanization and exclusion can be properly understood only through the study of local, provincial/state-level and national conditions. We thus conclude this section with a closer examination of peri-urban dynamics in a specific locale. The unusual density of high-quality research on the province of Cavite, near Manila, makes it a fruitful site for this purpose. Notably, there are three book-length studies: John McAndrew's *Urban Usurpation* (1994) and Philip Kelly's *Landscapes of Globalisation* (2000) draw on extensive ethnographic fieldwork in different decades, while John Sidel's *Capital, Coercion, and Crime* (1999) analyzes local and provincial politics in Cavite and Cebu. For much of the Spanish colonial period, Cavite's lowlands were largely controlled by friar estates, and by the late 19th century the province was producing rice for Manila and crops such as coffee and sugar cane for export. A highly unequal agrarian structure emerged on the estates, consisting of the estate owners, the *inquilinos* (non-cultivating tenants who leased land from the friars) and their sharecropping tenants. When the friar estates were sold off by the new American rulers early in the 20th century, the *inquilinos* and other large-scale buyers gained control of most of the land. During the US period agriculture continued to expand rapidly.[53] Smallholdings based on share tenancy rather than plantations were the rule during this period, but after World War II landless labourers came to make up an increasing percentage of the workforce.[54] The prevalence of tenancy means that for many of the farmers facing potential land conversion, the decision is not whether to give up their land but whether to give up their tenancy rights. Cavite has also long been a hub for illegal economic activity and is "notorious in Filipino popular culture" for banditry and political violence.[55]

The timing, extent and dynamics of peri-urbanization have varied within Cavite, and McAndrew and Kelly give a helpful sense of this diversity by comparing different areas. With respect to the "cutting edge" of peri-urbanization, however, we might (very roughly) divide the province's history into two periods, one beginning in the late 1960s and the second with the fall of Marcos in 1986. In his work on the municipality of General Trias, McAndrew traces Marcos-era efforts to establish industrial estates, agribusiness ventures and subdivisions. The municipality was attractive for these purposes in part because a state directive had earmarked land along a road running through General Trias for industrial purposes. Under this scheme, irrigated lowland areas were meant to be off limits to conversion (though in practice this was not always the case), while uplands were fair game.[56] McAndrew also notes of the General Trias village of Buenavista that while the percentage of households earning a living primarily from farming dropped between 1970 and 1983, the absolute number rose. Agriculture thus continued to expand even as livelihoods reoriented away from farming and as farms became worked more often by wage labourers rather than by tenants.[57] From around the mid-1980s (and earlier in some areas), however, the dynamics of peri-urbanization shifted and intensified as the Philippines began to move towards export-oriented industrialization and new models of residential development. Kelly demonstrates that the foreign direct investment that flowed into the Philippines from the late 1980s and, even more, the mid-1990s was spatially highly concentrated in Cavite and Laguna (another province in the Southern Tagalog region), and that as of 1995 virtually all of the industrial employment in Cavite was located in "the prime agricultural areas of the province's lowlands".[58] The 1990s also saw the rapid expansion of housing estates in Cavite. Some of these were for workers in the new industrial zones,[59] while others were aimed at more upscale commuters from Manila and incorporated golf courses and other land-intensive amenities. The concentration of industrial production in lowland Cavite ran counter to various government programmes and policies, including CARP, and Kelly writes that the most important aspect of government policy has been "the ineffectiveness and constant undermining of spatial planning, rather than its impact".[60]

Peri-urbanization in Cavite has taken place through a long series of land conversions, and Kelly, McAndrew and Sidel give an excellent sense of the range of these cases. Some of these land conversions were large-scale, others quite small; some were contentious, others quiet; some were (more or less) voluntary, others coercive or even violent; and a few cases became known nationally or even internationally. McAndrew's book *Urban Usurpation* begins with a discussion of the efforts of a joint venture between the state-owned National Development Corporation and Japan's Marubeni Corporation to convert 230 hectares of farmland in Langkaan, a village in the municipality of Dasmariñas, to an industrial estate. The struggle over whether this land would be exempted from CARP and approved for conversion pitted local farmers, peasant and cause-oriented groups, and the secretary of agrarian reform, Florencio "Butch" Abad, against the National Development Corporation, Marubeni, the Department of Trade and Industry, and the ruling party. The case became one of the most prominent disputes involving the new agrarian reform law, and by the time McAndrew's book went to press President Aquino had intervened, a Supreme Court case was in the offing, farmers who accepted compensation had reoccupied part of the property, and Abad had resigned.[61]

While McAndrew and Kelly report many contentious cases, and document intimidation and threats, they also draw our attention to more low-key moments of peri-urbanization. Examining the early development of residential subdivisions in General Trias, for instance, McAndrew writes that most "displaced tenants relinquished their rights without incident and accepted as the only measure of compensation a free home lot on the newly subdivided property". When protests became more common in later years, "the threat of a court case led [tenants] to accept compensation settlements".[62]

To understand the range of powers brought to bear in converting land or resisting its conversion, we must look at peri-urbanization from two perspectives: that of the people seeking to carry it out, and that of the people farming the land. A central aspect of the politics of land conversion in the Philippines is the highly decentralized system of land administration introduced by the Local Government Code of 1991.[63] As Sidel explains, elected municipal politicians in Cavite "control the awarding of building permits, the passage of municipal zoning ordinances, the use of government-owned land, the allocation of public works, the approval of reclamation projects, and, most important, the implementation of agrarian reform".[64] The fact that municipal governments receive more tax revenue from industrial and commercial than agricultural land uses gives them a strong incentive to use the power to zone in favour of conversion.[65] Likely more important, however, is the fact that many municipal and city governments in Cavite are controlled by local bosses who fuse real estate concerns with control of political office and the means of violence. Sidel writes evocatively that "[g]angster-style competition between rival bosses of the most predatory variety has long characterized small-town politics in Cavite, and elections in the province have resembled nothing if not the mobilization of rival armed camps for guerrilla warfare".[66] Indeed, politics is a rough enough game in the province that even elites are not secure from each other's predation, with landownership remaining, as of the late 1990s, "highly contingent upon the ebbs and flows of local politics".[67] These bosses have brought the extensive powers of office (including control over the police force and links to national patronage networks), their own wealth, and their gangs of "armed goons" to the twin projects of grabbing land for themselves and playing a brokering role in making land available to industrial and residential developments. Given that the bosses enjoy ideological and policy support at the national level,[68] it is clear that the balance is tilted firmly towards conversion in Cavite.

Turning now to an examination of the way in which powers of exclusion are used by, and act upon, the smallholders and tenants who farm the land, there is, as we point out in Chapter 7, a compelling story of mobilization. Many farmers have responded to the blatant capture of regulatory power by elites, and to the environment of threats and intimidation in which they find themselves, by making common cause with other farmers and with social movement organizations, occupying land, appealing to the law, and invoking discourses of equity and national food security. However, the mundane pressures that operate on an everyday basis to loosen people's hold on the land are probably more significant, though less dramatic. Kelly's fieldwork makes it clear that farmers in peri-urbanizing areas face great difficulty keeping up farming, for both economic and environmental reasons. On the economic side, the new employment opportunities created in Cavite and elsewhere by the rapidly expanding industrial and service sectors have made hiring wage labour more expensive and hiring it at harvest time more difficult, squeezing farmers and preventing them from moving into more labour-intensive high-value crops. The rapidly rising value of land has also, as discussed above, prompted landholders to remove their tenants for fear that the land may otherwise be brought under CARP.[69] On the environmental side, industrial and residential developments have interfered with irrigation systems and generated pollution, while idle land and garbage dumps breed pests that damage crops. These pressures on the viability of farming "make the offer of substantial disturbance compensation very attractive" for tenants.[70]

The power of legitimation can also push farmers to agree to conversion. One pressure is the difficulty that tenants have in bargaining aggressively with landlords who want to buy out their tenancy rights, given that their families have been involved in a hierarchical bond with them for generations (see Chapter 2). Kelly quotes one tenant to the effect that it would be "inappropriate" to "act superior to the owner of the land", and another who told him that "we just go along with the agreement because it's theirs, and it's inappropriate for us to say we don't want to. It will appear that we are becoming greedy over it."[71] Farmers are also more inclined to acquiesce to conversion when they feel, as most do, that it is both unlikely and undesirable that their children will become farmers. As livelihoods for Caviteños are increasingly to be found in industry and services, in Manila or overseas, farmers conclude that "[s]elling up their rights and using compensation money to pay for the completion of their children's education, or building homes for them in the village seem more realistic legacies for the next generation".[72]

Tourism

Tourism development and peri-urbanization share many features, and indeed in some cases are difficult to distinguish. Golf courses and other leisure facilities are common in peri-urban areas in Southeast Asia, and a third of Philippine courses were, as of 1999, located in the Calabarzon area around Manila.[73] The building of second homes in rural areas such as Hop Son, in northern Vietnam, and in scenic locations in the hills of northern Thailand also mixes aspects of peri-urbanization and tourism.[74] More broadly, both processes are components of deagrarianization. As tourism expands, local land, livelihoods and visions of self and of the future are reoriented away from agriculture. Further, tourism, like peri-urbanization, is profoundly connected to the global economy, not merely in the obvious sense that many tourists are foreigners, but also through the involvement of overseas developers, designers, managers, capital and labour, and through the support the sector has received from foreign aid. Tourism is promoted by a familiar combination of state actors (who rezone land, create tourism master plans and promotional campaigns, build infrastructure, and try to deal with resistance) and powerful corporate actors, both domestic and foreign. Our emphasis here is on the extent to which powers of exclusion have been mobilized in similar ways in both peri-urban and tourist settings.

As in the peri-urban case, perhaps the best starting point is rising demand for land and the implications, welcome and unwelcome, that it has for land-holders. The dynamics vary from place to place. Peter Williamson and Philip Hirsch show how tourism's implications for land relations on the Thai island of Koh Samui depended on the prior history of land accumulation and sale. Before tourism, when coconuts were virtually the only agricultural product on the island, Sino-Thai traders were more inclined than ethnic Thai to accumulate land, including seemingly non-productive beach land, for its own sake. When ethnic Thai went into debt to these traders, they were likely to cede beaches before giving up agricultural land. As a result, when the tourism boom began in the 1970s and outside capital poured into the local land market, it was this Sino-Thai elite that reaped windfall profits, both directly and through working as land agents or brokers.[75] Dararat Kaewkuntee's work on tourist areas on the Andaman Sea coast of Thailand, on the other hand, points to a different set of impacts of bubbling land markets on smalholders.[76] Kaewkuntee reports that villagers have been prompted to sell their land both by positive price incentives and by the difficulties of continuing to farm as the land around them is converted to tourist uses, and that having sold their land many people end up in low-paying jobs. This finding resonates with the peri-urban research of Kelly and To Xuan Phuc. In a variety of ways, then, the "purely economic" opportunities created by tourism have rearranged the distribution of land.

As one might expect, smallholders in tourist zones face not only the same opportunities as do those in peri-urban areas, but the same risk: the rising land prices that hold out the promise of gain also prompt others to lay claim to their land through non-market powers. As with peri-urbanization, while the power to regulate land use is certainly important, what stands out is the ability of powerful actors to subvert, evade or simply rewrite these regulations. A study on golf courses by the Philippine Human Rights Information Center, for instance, tells a familiar story of ineffective national-level efforts at regulation, capture of zoning at the local level, land that had been brought under CARP being declared exempt from reform, and displaced farmers.[77] *The Golf War*, a 1999 documentary on the Harbortown resort project under construction in Hacienda Looc, graphically illustrates all of these processes; it also shows the extensive use of force and intimidation (including murder) on behalf of the project by security guards and police, and the eventual response of the New People's Army.[78] Post-tsunami events along Thailand's Andaman Coast represent an unusual, and perhaps unusually distressing, example of force being used to lay claim to land for tourism. When the tsunami killed many coastal residents and destroyed buildings, landmarks and (in some cases) title documents, it created, if not a blank slate, at least a particularly hospitable environment for land-grabbing.[79] The state, too, used arguments about safety to exclude people from returning to coastal land.[80]

As with peri-urbanization, the dynamics of exclusion surrounding tourist development have involved a wide range of actors and have played out at a variety of scales. Carol Warren's research in Bali shows how villagers have sought to exclude outsiders, and each other, from access to land. Across Indonesia, the regional autonomy/decentralization legislation passed in 1999, in conjunction with the increased scope for resisting state and corporate action after the fall of Suharto, led to efforts by villages to improve the terms of their relationships with state and private actors. On Bali, she writes, "local communities … have been assuming management of tourist attractions, developing new industries, regulating resource access, and negotiating terms with hotels and other outside interests to achieve higher levels of local employment and funding of community programmes."[81] One element of this process has been NGO-supported community mapping projects (see Chapter 3). Comparing across five "rural, remote, and [...] more or less ethnically homogeneous" communities that took on such projects, Warren encountered strong interest in tourism's economic opportunities, but also conflict with regional or provincial authorities over land, water and/or tourism development.[82] She also found significant differences among the sites.

As mapping interacted with pre-existing priorities, social divisions and interpersonal conflicts, the consequences of efforts to identify and assert control over resources varied greatly. In Ceningan, villagers reversed their initially positive stance towards government plans for tourism development and rejected a proposed US$200 million resort. In Belok Sidan, however, divisions over how best to assert local control over an existing tourism development generated enough tension that the village withdrew from the mapping programme.[83] Indeed, while the mapping projects were viewed positively by most participants, they also stimulated contention as explicit questions about the conditions of resource access were raised. One of the core issues was the relationship between the right to buy and sell land, and membership in the community. Villages across Bali have been "involved in intense debates on the need to introduce or reinforce customary rules which would permit the sale of village land only on condition that the purchaser reside in the village, join the *banjar/desa adat*, and take on full community and ritual service obligations".[84] The mapping exercise also put one of the core conundrums of land exclusion front and centre: that is, it "brought to a head fundamental tensions between land as a symbol and cultural resource on the one hand, and as a productive commodity in market-oriented economic development agendas on the other". One of Warren's informants gave a pithy account of these discussions:

> The debate was really intense, from one side: 'This is my land, why can't I develop it? Why is the *banjar* prohibiting sale?' 'The reason is like this—for the environment, for the next generation ...' 'Well, if that's the reason!' 'Do you agree?' 'Agreed!'[85]

While Warren's work in Bali primarily highlights the local, provincial and national scales, Tim Winter's research on Angkor, a UNESCO World Heritage Site since 1992, shows regional and global actors becoming involved in struggles over land for tourism development. Tourism at Angkor exploded during the 1990s: in 1993 the site sold around 9,000 tickets, but by the early 21st century annual sales were closer to 750,000, with much of the demand coming from Japanese, Chinese, Taiwanese and South Korean tourists.[86] The International Coordinating Committee for the Safeguarding and Development of Angkor, created under UNESCO, saw this rapid expansion primarily "as a threat, and a danger to be contained".[87] An international group of cartographers, historians, ecologists, hydrologists and other experts put together a management plan, and on the basis of this plan a Royal Decree authorized the Zoning and Environmental Management Plan of 1994. This plan divided the Angkor site and the neighbouring town of Siem Reap into zones for protection and development, and included plans for a hotel zone outside the park, in order to protect the site from degradation. A UNESCO-created, Cambodian-run management body—the Authority for the Protection and Safeguarding of the Angkor Region—organized the legal side of the hotel zone, purchased the land, and in the late 1990s relocated thousands of people living in the Angkor area. The hotel zone scheme, however, went nowhere: by early 2005, "rather than being a space of luxury hotels and landscaped gardens, the zone [was] little more than a grid of roads and kilometers of wire fencing".[88] Part of the reason for this failure, Winter argues, was the priority French experts in the International Coordinating Committee placed on finding European or North American hotel chains to take part, in order to promote what was referred to as "high quality, cultural tourism" rather than the mass variety.[89]

In contrast to these failed efforts, developers targeting mid-range package tourists from Northeast Asia were able very quickly to create a highly successful strip of restaurants, shops and hotels along the highway linking Angkor to Siem Reap International Airport. By the late 1990s, Korean, Japanese and Chinese operators and capital had entered the area in significant numbers. This boom area was created through "an extremely high level of entrepreneurialism"[90] and a willingness to ignore regulations. Getting around the rules was facilitated by the fact that much of the land in question had been awarded to high-ranking Cambodian military officials as part of the post-war demobilization process. Tourism development in the Angkor area, then, has seen a collision between two dramatically different visions: one oriented to zoning, conservation and Western expertise, concerned with making sure that "politics and legislation [are] as rigid as possible",[91] and largely a failure; the other connected to Northeast Asia, highly market-driven, associated with the military and conspicuously successful.

Dams

We conclude this chapter by turning to infrastructure. Roads and highways, ports and canals, airports, power plants, dams, and waste management facilities all require land, and large amounts of agricultural land have been converted to these uses in Southeast Asia. Infrastructure projects almost always require the exclusion of former land users, and one of the arguments of this section is that these projects are more likely than land conversion in peri-urban and tourist settings to involve involuntary displacement.[92] Large-scale dams are among the largest infrastructural projects undertaken in Southeast Asia (or anywhere), and by focusing on them in this section we bring the exclusions associated with infrastructure into sharp relief (we have also highlighted some of the problems associated with dams in our discussion of the Nam Theun 2 project in Chapter 3). Large dams are enormously expensive, can take more than a decade to complete, and are close to the heart of a certain vision of economic progress and national prestige, and for these reasons have been referred to as "temples of development" ever since Jawaharlal Nehru referred to them as the "temples of modern India". They also have dramatic consequences for people living both upstream and down. Upstream, the newly created reservoir can inundate hundreds of square kilometres of land; downstream, changes in the flow of the river can destroy livelihoods. The consequences of large dams in Southeast Asia have been felt directly by hundreds of thousands of people, and indirectly by millions more.

Struggles over dams also mobilize a wide variety of actors at a range of scales. The people affected by dams, of course, have mobilized against them in a number of ways that we examine closely in Chapter 7. Representatives of different branches of the state take an interest, though they do not necessarily all push in the same direction.[93] What stands out in comparison with most of the peri-urban and tourism developments presented above, however, is the ubiquitous involvement of transnational actors, both regional and global, in the politics of Southeast Asian dams. The Mekong River Commission, which was created in 1995 and has representatives from Cambodia, Laos, Thailand and Vietnam, has attempted to guide dam development within the Mekong River Basin.[94] The funding and loan guarantees of the Asian Development Bank and the World Bank, as well as various national development agencies, have been indispensable to the construction of many of the region's large dams, and the planning and reports undertaken by the staff of multilateral development banks and a small army of consultant hydrologists, biologists, anthropologists, and other physical and social scientists have shaped dam construction.[95] Local, national and international NGOs, too, have thrown themselves into the struggle over dams, generally on the opposing side but in some cases—that, for instance, of International Union for Conservation of Nature's endorsement of Nam Theun 2—in support. The mobilizations that have taken place over some of the highest-profile dam projects—Chico River in the Philippines, Nam Choan and Pak Mun in Thailand, Kedung Ombo in Indonesia, Bakun in Malaysia—have become touchstones in debates about Southeast Asian civil society (see Chapter 7).[96]

The exclusionary effects of dams can be divided into two main categories. Most directly, exclusion takes place when land is submerged by the dam's reservoir. The people who face the inundation of their land rarely (if ever) view this prospect as representing any kind of dilemma. While landholders in areas being developed for peri-urban or tourist uses are often keen to sell their land in response to market demand, those in future reservoir areas face instead a state claim that will result in the flooding of their land, the dismantling of their community and their (often involuntary) resettlement. The issue here is thus not rising land values but rising waters. The power of prices does enter this process through state offers of compensation for lost assets and relocation expenses. However, these offers are almost always insufficient to prompt anyone to move absent the threat of the dam, and sometimes no offer is made. In some areas, this is so because the state does not recognize the claims to the land made by its current inhabitants—for instance, when they are living on land claimed by the state as political forest. The people to be relocated are also often from ethnic minority groups whose interests may not be given high priority by national states.[97] People relocated to other agricultural areas almost always experience a substantial (if not shocking) decline in their standard of living. Regulation, force and legitimation thus stand out in studies of large-scale dam development as the relevant powers of exclusion far more than does the market.

However, market demand for electricity plays a large role in driving dam construction, quite starkly so when the energy is sold across national borders and the dams are privately owned commercial operations.

Dams create exclusion well beyond the reservoir. We have already noted the expansive effect of dam-related conservation initiatives (see Chapter 3). The environmental impacts of river flow changes can make it much more difficult for people both upstream and down to access their land. Hirsch and Andrew Wyatt demonstrate this in their study of the implications for downstream people in Cambodia of the construction in Vietnam of the Yali Falls Dam on the Se San River, a tributary of the Mekong. Most dramatically, they report that unpredictable floods caused by manipulation of water levels in the dam have caused at least 39 deaths, in addition to washing away livestock, boats, fishing nets and crops. These floods have made riverbank agriculture almost impossible and have prompted some villagers to move away from the river (and its benefits that drew them there in the first place). Impacts on water quality and fisheries have also had severe consequences for affected communities.[98] Dams also generate exclusionary dynamics when the people who have been relocated come into conflict over resources and land with people already resident in the areas in which they are trying to settle. These impacts are discussed in the study of Hoa Binh below.

While all the powers of exclusion are on display in the politics of dam construction in Southeast Asia, it is worth dwelling for a moment on struggles over legitimation as they apply to dams. Like the large-scale conservation-oriented enclosures we examined in Chapter 3, large dams have generated highly charged debates over the principles guiding land use, and indeed over the common good. This is to some degree because of the enormous scale of large dam projects in terms of area inundated, numbers of people relocated or otherwise affected, changes in river flow, money spent, debt incurred and power generated. Dam proponents such as state actors, corporations, development banks and aid organizations, engineers, and scientists promote these projects for three functional reasons: electricity, irrigation and flood control. For large dams, the electricity rationale is by far the most important. These functions are promoted in terms of the benefits they will bring the nation. Often, they are also phrased in terms of impending crisis: if the dam is not built, electricity generation will be insufficient to keep up with growth, farming will suffer, and floods will continue to devastate communities along the river. From either perspective, opponents of dams can be painted as enemies of progress. Perhaps even more profoundly, however, dams have a powerful symbolic value as a representation of national development and technological prowess that transcends their functional value. George Aditjondro provides an excellent example of this type of thinking in his discussion of the career of Sutami, the first public works minister under Suharto:

> As a member of the generation that had fought for freedom, Sutami often emphasized that Indonesian engineers were facing a new form of *perjuangan* (struggle) to prove their technical expertise, which had often been ridiculed by colonial engineers. ... He challenged his younger colleagues to put the *perjuangan* (struggle) spirit to work by building as many dams as possible, even with very limited equipment.[99]

In some cases, dam proponents simply do not accept that dams bring costs as well as benefits. In others, they acknowledge that there are negative consequences for some, but see them as a price that has to be paid in order for development to occur. According to this logic, local opponents of a dam are "selfish" people who are not willing to sacrifice their own interests for the greater good of the nation. As government officials tried to persuade villagers to accept their relocation from the vicinity of the proposed Tab Salao Dam in Thailand in the 1980s, for instance, they asserted "that local people were selfishly obstructing a project with widespread benefits in order to pursue their own narrow interests".[100] It is at this intersection of (purported) national benefit and local sacrifice that much of the critique of dams has been located. At times this critique extends to opposition to the construction of the dam in question, but it can also be focused on the compensation that was promised, or that seems just, but has failed to materialize. The struggle over legitimation can also be framed, however, in terms of a rejection of the idea that dams are in fact beneficial for the nation. Given a long history of cost overruns in dam construction, the massive debts that Southeast Asian states have taken on in order to build them, the tendency of planners not to ask if there might be cheaper and less destructive ways to generate electricity, and the persistent failures of dams to generate as much power as their proponents claimed they would, it is not surprising that some opponents argue that the opposition of local sacrifice to national gain is a false one.[101] This line of argument suggests that another important reason dams are built is that they are massively beneficial to the well-connected construction and engineering companies (domestic and foreign) that build them, to the international development banks that finance them, and to the army of social, economic and environmental "expert" consultants who plan for and evaluate their wider impacts.

Hoa Binh

By the early 21st century, efforts to build dams in most of Southeast Asia confronted an environment of close scrutiny by local, national and transnational opponents. International financial institutions such as the World Bank and ADB, having been burned by well-coordinated transnational anti-dam campaigns, had become much more picky about supporting dams. Construction of Vietnam's Hoa Binh Dam started in the 1970s under very different conditions. The dam, located on the Black River (Song Da) about 75 kilometres west of Hanoi, was a massive undertaking. It took 12 years to build, cost around US$1.5 billion, created a reservoir 230 kilometres long covering 200 square kilometres, flooded 11,000 hectares of prime agricultural land, displaced 58,000 people from 20 different ethnic groups, and on its completion was the largest hydroelectric project in Southeast Asia, with a maximum capacity of 1,920 megawatts.[102] Completed in 1991, the dam still generated roughly 40 per cent of all Vietnam's electricity by the early 2000s, and 15 per cent by 2010, so its importance to the national economy is difficult to overstate.[103] The dam was constructed with aid not from the World Bank but from the Soviet Union, and while Vietnamese economic policy changed rapidly during the final years of the dam's construction, the resettlement programmes undertaken bore a clear socialist-era stamp in being based on a collective farming model. Until the 1993 changes to the Land Law (see Appendix), resettlement "was regarded as a simple compensation issue" of moving people to a new location.[104] The resources devoted to this project, however, were risibly low and were generally inadequate to cover even moving expenses, let alone to replace the value of lost assets or to help create new sources of livelihood.[105] Little thought was put into planning for resettlement, and the mechanics were left to local governments that were unprepared for the challenges entailed.[106]

A study carried out just after the completion of the dam by Hirsch and his colleagues reports that the evacuees, almost all of whom were wet rice farmers, responded to their exclusion from the future reservoir site in two ways. The minority, roughly one quarter of the affected people, relocated to official resettlement sites. The majority simply moved up from their lowland homes and land into nearby hills, and shifted from wet rice to dryland farming. Many of these people, in fact, moved several times, partly because they could not believe that the level of the reservoir was going to rise as high as the state claimed it would, and partly because they needed to stay close to the water for transportation and other purposes. Many of those who moved to the official resettlement centres ended up returning to the areas above the reservoir.[107] As people tried to make a living in the hills, they rapidly cleared the forests, increasing erosion and, through siltation, significantly reducing the projected life of the dam. Those who did try to make a go of it at the resettlement locations, on the other hand, faced conflicts over land and resources with the people who already lived in those areas, and, again, were forced into unsustainable livelihood practices. Indeed, as Hirsch and Bach Tan Sinh note, increased pressure on resources and conflict between old and new residents contributed both to the breakdown of systems of control over resource access, and to efforts to demarcate more clearly which forest belonged to which community. The process of relocation also accelerated the collapse of collective farming, as relocated families found that they had to fend for themselves.[108] Ambiguities over tenure, created in the gaps between a resettlement process that assumed a collective farming model and the realities of the situation, contributed to conflicts over land.[109] Virtually all of the evacuees experienced a dramatic drop in their standard of living. They also felt "a commonly expressed sense of grievance at having sacrificed their lands and livelihoods for the wider community, for national development, and having received very little in return".[110]

Conclusion

In this chapter, we have explored the major transformations in land access and exclusion stimulated by post-agrarian land uses across Southeast Asia. Demand for land for industrial, residential, commercial and tourist uses, and for the infrastructure to support those uses, has created new incentives for actors to mobilize the powers of exclusion at their disposal to make claims to land. Our analysis makes it clear that the farmer response to these incentives cannot be understood solely in terms of resistance or defence of peasant livelihoods. Again and again, smallholders and tenants have, by their own choice, taken up the offers of payment made to them and relinquished their claims to land. We also show, however, that these choices are made in the context of the use of regulatory, market, coercive and legitimating power by other actors. Even smallholders in peri-urbanizing Southeast Asia who wish to sell up generally do so in market conditions biased against them by regulatory decisions and by unequal market power. Farmers trying to hold on to their land in the midst of a tourist development often face the very real threat of force. And while smallholders have been able to invoke discourses of equality, rule of law, citizenship and food security in defending their claims to land against dams, new towns and resorts, we have found that agriculture almost always ends up on the back foot in debates over the kinds of land use that are most desirable.

Two of our findings in particular bear repeating here. First, this chapter highlights one of the points made in Chapter 1: While it is useful to separate our four powers of exclusion for analytical purposes, in practice they are interwoven. When land is being converted to peri-urban and tourist uses, people with market power—those able to offer a relatively high price for the land in question—often also have access to the powers of regulation and coercion. This fusion of the economic and the political has profound implications for the functioning of land markets and for the dynamics of exclusion. Second, and relatedly, it is striking that the areas of agricultural land that seem to be at the cutting edge of interaction with more "modern", "globalized" processes such industrialization, urbanization and tourist development often end up looking like "Wild West" frontiers. In particular, land-grabbing—usually associated with lawless, distant and "underdeveloped" areas—has been a central part of these processes.

Notes

1. Rigg 2001. See also Rigg 2006.
2. Rigg 2001.
3. Askew 2003.
4. This word is a compound of the Indonesian words for "village" and "town". See McGee 1989.
5. Kelly 1999: 285; Dick and Rimmer 1998; Webster 2002: 5–6.
6. Jabotabek and Calabarzon are nicknames composed from the names of the provinces they are made up of: Jakarta, Bogor, Tangerang and Bekasi for the former, and Cavite, Laguna, Batangas, Rizal and Quezon for Calabarzon.
7. A fascinating series of maps presented by Murakami and his co-authors gives a sense both of the heterogeneity of peri-urban land use and of the sheer size of some of these extended metropolitan regions: Murakami *et al.* 2005: 257.
8. Kelly 2003: 175–6. See also Maneepong and Webster 2008: 143.
9. There is some debate over whether peri-urbanization is peculiarly Southeast Asian. McGee 1991, Shatkin 2000 and Murakami *et al.* 2005 argue that it is, while Dick and Rimmer 1998 disagree.
10. Askew 2003: 288.
11. Webster 2002. See also Kelly 2000: Chapter 3.
12. Webster 2002: 9–10; on Jabotabek, see Goldblum and Wong 2000: 32; see also Kelly 1999 for Manila and Firman 2002: 233 for Indonesia.
13. Kerkvliet 2006: 297–9.
14. For Thailand, see Glassman and Sneddon 2003.
15. On the Philippines, see Kelly 2003: 180–1; Sajor 2003: 729; on Thailand, see Maneepong and Webster 2008: 139–40. For the Indonesian situation and a brief discussion of potential implications for urban planning, see Firman 2002: 237–40.
16. Firman 2000: 16.
17. Askew 2003: 306.
18. Firman 1997: 1042; Kelly 2003: 183.
19. Firman 1997: 1042.
20. Askew 2003: 287.
21. Askew 2003: 301.
22. Askew 2003: 310.
23. To 2007, unpublished work cited with permission.

24. Kelly 2003: 177.
25. The difficulty of assembling these plots has inhibited large-scale developments around Bangkok (Webster 2002: 16; Askew 2003: 307).
26. Ferguson and Hoffman 1993: 62.
27. To 2007: 13.
28. Ferguson and Hoffman 1993: 57; Leaf 1994: 344.
29. Firman 1997: 1041; Leaf 1994: 345.
30. Ferguson and Hoffman 1993: 59.
31. Kelly 2003: 179–181.
32. Kelly 1999: 299; Malaque and Yokohari 2007: 193. See also Firman 1997: 1041.
33. This tendency predates CARP. See McAndrew 1994: 122, 126–31, 170.
34. Kelly 1999, 2003; Sajor 2003; Sidel 1999.
35. Sajor 2003: 732.
36. Leaf 2002: 27.
37. Kelly 2003: 183. See also Lange 2010: 62–5.
38. Firman 2000: 16.
39. Sajor 2003: 734; Webster 2002: 6.
40. Ferguson and Hoffman 1993: 65.
41. Cited in Firman 2000: 15.
42. Ferguson and Hoffman 1993: 63.
43. Sombat 2000: 61–2; Shatkin 2004.
44. Dick and Rimmer 1998: 2317. On the culture of the Jakarta suburb and its role in national status, see also Cowherd and Heikkila 2002.
45. Kerkvliet 2006: 299. See also Labbé 2010.
46. Leaf 1994: 345, 348, 349.
47. Leaf 2002: 28.
48. Rigg and Salamanca 2006: 14.
49. Askew 2003: 289.
50. Firman 2004: 363; Shatkin 2008.
51. Sidel 1999, Shatkin 2008.
52. Shatkin 2008: 12. See also Anderson 1998.
53. McAndrew 1994: 18, 25, 27, 35, 37.
54. Sidel 1999: 21.
55. Kelly 2000: 68; for details see McAndrew 1994: 28–32, 42–5.
56. McAndrew 1994: 117, 115.
57. McAndrew 1994: 96–100.
58. Kelly 2000: 57–60.
59. Kelly 2003: 175.
60. Kelly 2000: 63–6.
61. McAndrew 1994: xi–xiv.
62. McAndrew 1994: 128.
63. Kelly 2003: 180–1; Sajor 2003: 729.
64. Sidel 1999: 33.
65. Malaque and Yokohari 2007: 193.
66. Sidel 1999: 28.

67. Sidel 1999: 32, 47.
68. See, for instance, McAndrew 1994: 117.
69. Kelly 1999: 293–4.
70. Kelly 1999: 295–7; quotation at 296.
71. Kelly 2003: 185. See also McAndrew 1994: 131–4, 170.
72. Kelly 1999: 298.
73. Tulod-Peteros 1999: 3.
74. Forsyth 1995: 882.
75. Williamson and Hirsch 1996: 191–2.
76. Kaewkuntee 2006.
77. Tulod-Peteros 1999: 52–8, 66, 77.
78. On the film, see www.golfwar.org. See also Tulod-Peteros 1999: 45–6, 52–4, 69–70, 88–90, 103; Cole 2002.
79. Kaewkuntee's work in Phang Nga province provides examples of land-grabbing by private companies and gives a sense of the very different trajectories of land conflict in four affected communities (Kaewkuntee 2006).
80. Fisher 2006.
81. Warren 2005: 51–2.
82. Warren 2005: 54.
83. Warren 2005: 57–8.
84. Warren 2005: 62.
85. Warren 2005: 62, 63.
86. Winter 2007: 28.
87. Winter 2007: 33.
88. Winter 2007: 33–5.
89. Winter 2007: 35–6.
90. Winter 2007: 37.
91. Winter 2007: 36.
92. Extensive infrastructure has been built in peri-urbanizing and tourist areas, and conversion for infrastructure is thus not entirely separate from the other topics covered in this chapter.
93. See, for instance, Hirsch 1998b: 62.
94. See Mitchell 1998; Bakker 1999; Lebel, Garden and Imamura 2005; Hirsch 2010.
95. Goldman 2005.
96. On the campaign against the Chico River Dam (which was never built), see Morales-Fernholz 2002; on Nam Choan, Rigg 1991; on Pak Mun, Fahn 2003; Kuhonta 2009; on Kedung Ombo, Aditjondro 1998; on the Bakun Dam, Yee 2005.
97. Bakker 1999: 218.
98. Hirsch and Wyatt 2004: 56–8.
99. Aditjondro 1998: 32.
100. Hirsch 1998b: 61.
101. On the gap between the promise and reality of Thailand's Pak Mun Dam, see Hirsch 2010: 315.
102. Hirsch and Sinh 1992: 8–9.

103. Yen 2003: 17; *Thanh Nien News*, 21 June 2010.
104. Yen 2003: 11.
105. Hirsch and Sinh 1992: 12.
106. Yen 2003: 29.
107. Hirsch and Sinh 1992: 14.
108. Hirsch and Sinh 1992: 15, 18; Hirsch 1998b: 65–7.
109. Hirsch 1998b: 65.
110. Hirsch and Sinh 1992: 16.

6

Intimate Exclusions: Everyday Accumulation and Dispossession

The process we explore in this chapter is one in which social intimates exclude one another from access to land as part of a strategy to accumulate capital. We highlight the ways in which processes of accumulation and dispossession work at close quarters, among neighbours and kin who share common histories and social interaction. For the most part, these are "everyday" processes, mundane and piecemeal, that do not grab headlines. But cumulatively, they have the effect of producing agrarian classes with differential access to means of production.

Agrarian differentiation is long-standing in Southeast Asia, especially in the more densely settled agricultural cores. In these areas, distinct classes of landlords, tenants and landless workers have existed for at least one century or sometimes two, and smallholders routinely exclude kin and neighbours from access to land, idioms of "moral economy" or "shared poverty" notwithstanding. To varying degrees, it has also been a characteristic of land relations in the more sparsely settled uplands, especially in the context of crop booms that make land especially valuable, and locally scarce. It merits renewed attention for two reasons. First, the smallholder-driven crop booms we described in Chapter 4 are expanding their spatial scope, as frontier areas are brought under intensive cultivation, and competition for land among "locals" and between locals and migrants is producing tension and conflict in many regions.[1] Second, the closing of forest frontiers and other processes that reduce smallholder access to new farmland—processes we described in earlier chapters—have made it increasingly difficult for people who lose land through processes of exclusion in one location to move off to try again in another.

How, then, does intimate exclusion come about? Practices that enable some people to accumulate land and capital at the expense of their neighbours and kin include setting a price for land purchase or rent that the latter cannot afford; lending them money or food at rates they cannot repay; renting or buying their land at "fire sale" prices; enclosing village commons so they cannot use it; and finally, distributing development handouts (cheap credit, free inputs, licences and access to land) to households that already have a competitive advantage (often office holders, entrepreneurs, major landholders or "progressive farmers"), increasing their capacity to accumulate land and capital for themselves and exclude others.

The latter two practices (enclosure of common land and uneven distribution of development benefits) draw most directly on powers of legitimation, specifically the need to increase productivity, although force and regulation also play a role. The first three draw on market powers, as they involve prices for rent or credit. Market competition is limited, however, when just a few landlords or local moneylenders hold a monopoly, and multi-stranded dependence on patrons makes it risky to alienate them. More generally, in intimate settings, market powers always need to be hedged around by attempts at legitimation, because social proximity freights exclusion with moral weight and has consequences for personal standing. Even when markets, and market prices, are well-established, it is hard for a woman to say to her sister, "This land is mine; you can use it only if you can buy it or rent it at the market price." Nor can a village patron, offered an attractive price for his land by someone from the city, easily evict a client who has rented the land for decades. In each case, further arguments are required to justify applying a market calculus, or a legal precept such as title, in contexts where other legitimating discourses (responsibility, kinship, friendship, custom, status, belonging, need) still have persuasive power. In periods of transition, when new forms of exclusion have yet to become routine, exclusion among intimates signals a wrenching reconfiguration of social bonds. Needless to say, the tension produced by exclusion's double edge is intense at an intimate scale, as each villager's assertion of a right to exclude runs up against another villager's claim for access. Yet it is difficult to find a Southeast Asian villager who does not want a piece of land to call his or her own.

Our analysis in this chapter builds on a long tradition of research on agrarian differentiation influenced by Marx and Lenin, one that examines processes of accumulation and dispossession among villagers in the context of the emergence of agrarian capitalism.[2] Scholars working in this tradition point out that accumulation becomes systematic only once the market shifts from being a site of opportunity (the opportunity to buy or sell for a profit) to one of compulsion (see Chapter 1).[3] In Southeast Asia, as in Europe, farmers have been buying and selling crops for millennia, yet for most of this time they have not been *compelled* to buy or to sell if they found the price unattractive. Landless people, however, have no choice—they must sell their labour. Similarly, households with insufficient land to provide food for home consumption are compelled to intensify production to produce higher-value crops for the market, and use the proceeds to buy food and farm inputs. Low prices for their crops, or high prices for farm inputs, put them at risk of the "simple reproduction squeeze", in which they enter a downward cascade that may, sooner or later, push them out of production and force them to rent out or sell their land.[4]

Our approach to agrarian differentiation pays particular attention to the market powers that work to make accumulation among intimates systematic and widespread, but we stress that this process does not proceed on autopilot. As we will show in the pages to follow, markets are mediated by social calibrations of many kinds, and discourses that attempt to legitimate exclusion are routinely contested. Even though decades of empirical research in rural Southeast Asia have confirmed that the prevalence of "shared poverty", a "moral economy" and the obligation to supply a "subsistence guarantee" have been exaggerated, the opposing argument that a market calculus, and market-based powers of exclusion, operate in a social vacuum is also problematic.[5] It takes human agency—socially situated practice—to create and sustain the conditions necessary for a market to operate, and to insinuate "the market" into intimate relations to the point where it overrides other considerations.

We begin with an examination of exclusion and accumulation in village Java, the site of classic debates about whether co-villagers can be said to "share poverty" by dividing land into tiny parcels and offering tenancy, sharecropping and work-distributing arrangements designed to ensure that everyone has access to an agrarian livelihood (as Clifford Geertz famously argued) or, alternatively, whether they exclude each other from access.[6] A study by Jonathan Pincus that traced distinct patterns of exclusion in three adjacent villages in Java's rice-producing core enables us to move beyond the static either-or, to specify more clearly the powers deployed as landholders and workers struggled over agrarian resources in a context where they were not strangers, but neighbours and kin.

Our second study concerns intimate exclusion by means of the private, individual enclosure of common and collectively inherited land in upland Sulawesi, a process tracked by Li since 1990. In this case "enclosure from below" was followed by a rapid process of agrarian differentiation, as the enclosed land quickly became a commodity accumulated by some households, to the exclusion of kin and neighbours caught in the "simple reproduction squeeze" and obliged to sell up. By 2009, the exclusion of intimates from access to agricultural land—a condition unthinkable in 1990—was taken for granted as a "normal" fact of highland life. We explore the powers deployed by the actors who brought about this transformation, the struggles that unfolded at different stages, and the discourses deployed by highlanders at the winning and losing ends of this trajectory.

Finally, we turn to Vietnam to examine how regulations put in place since 1988 to legitimate the private use and transfer of land have enabled a re-emergence of agrarian differentiation, a process proceeding rapidly in some villages—but not where villagers acted collectively to put countermeasures in place. We also examine how the uneven distribution of "development" inputs (roads, seeds, subsidized credit) and access to forestland have further entrenched emergent inequalities. This phenomenon is not unique to Vietnam, but the combination of authoritarian rule and the rapid emergence of a land market have made it especially prominent in the cases we examine. Here, as in all our examples, our focus is on the powers through which conditions for "intimate exclusion" are established and contested.

Shared Poverty in Java Revisited

The weight of the historical and ethnographic work published since Geertz described "shared poverty" on Java confirms that he overstated the existence of village-level mechanisms to even out access to land.[7] Already in the 19th century, the exclusion of co-villagers from access was common. Land was individually owned, and many villagers were landless;[8] landholders charged market rents; and moneylenders charged high interest, making mortgage and debt important routes to land accumulation, on one side, and distress sales on the other. Dutch observers around 1800 were already alarmed by the rate of landlessness resulting from the sale of land to cover debt, a problem they attributed to the shortsightedness of natives on one side, and the greed of Javanese and Chinese merchants on the other. Merchants who were not "intimates" could charge interest of hundreds or even thousands of percent, but even among neighbours and kin, interest of 50 per cent was common.[9] Landless people were incorporated into farm households as servants, or roamed the countryside in groups undertaking seasonal work. While some sought new land on forest frontiers, these settlements tended to replicate the hierarchy of longer-established areas because new arrivals with no capital had to buy food on credit, or work as unpaid labour, while waiting for their first crops to yield. The best land was claimed by the "founding" families, who established themselves as village officials, collected taxes and offered protection.[10] Direct control over labour through debt and its stronger variant, debt slavery, was a common means of accumulation at this time, especially in conditions where land was abundant, making it difficult to exclude people from access to it.[11]

Colonial concern about high levels of debt and landlessness among the rural population led to measures to combat usury and forbid the sale of land, but other colonial policies exacerbated these problems by starving smallholders of capital through excessive taxation, and consolidating the position of the larger landholders.[12] By the end of the 19th century, very few rural households on Java could survive on their own outside market relations. They were compelled to buy food, rent land in or out, take on debt, sell labour, or sell produce. They were subject to intensified demand from the colonial state and its proxies, and to the vagaries of world demand for commercial crops, a fact that hit especially hard during the 1930s depression.[13]

Researchers have shown that the small pockets of communally owned village land that still existed in Java in the 1950s did not represent the persistence of a shared commitment to livelihood security for the poor, as Geertz proposed. Rather, this "communal" land was a residue of the colonial cultivation system that reshaped agrarian relations in the period 1830–70. Communal land further dwindled in significance as it was privatized and appropriated by village elites in subsequent decades. Sharecropping and work-sharing arrangements were shaped by an economic calculus in which landowners used their capacity to grant access to land, work and credit as a strategy to exploit and discipline labour. They were also used reciprocally among kin and neighbours from similar economic strata as a means to mitigate risk by spreading the range of land types, seed varieties and harvest timing. The poor had their own circuits of exchange, as kin and neighbours made tiny loans to each other in times of crisis, with the expectation of a similar return. Significantly, there were no cross-class mechanisms for sharing among villagers in ways that guaranteed access to land as a basis of livelihood, or otherwise interrupted processes of accumulation in order to prioritize the survival of the poor.[14] Further, the situation in Java's villages in the late 1950s and early 1960s was anything but harmonious: two years after Geertz published his book, up to half a million landless and land-poor villagers with alleged Communist affiliations were massacred by the army and youth groups, organized as militia. These actions were supported by rural landlords who were afraid that the mobilized masses would take over their land.[15]

The use of rental, sharecropping and credit relations as mechanisms of extraction was powerfully illustrated in Frans Hüsken's study of a village on Java's north coast. Hüsken used the term "rent capitalism" to describe a form of accumulation in which a person supplying credit extracts the surplus product and labour of smallholders *without* buying up their land.[16] He found two versions of this practice. In one version, farmers retained control of their land but lost access to any surplus beyond bare subsistence through rates of interest on production and pre-harvest loans of around 50 per cent. They were, in effect, disguised wage-workers, with the disadvantage that they had to bear the costs and risks of production. In the second version, farmers were obliged to rent out their land to larger landowners who had advanced credit, with the rental terms operating as a disguised form of usury. Some were re-employed to work their own land as sharecroppers. The children of indebted villagers and sharecroppers worked for their patrons without pay as servants and farm-hands. This was a village in which about 5 per cent of the households owned 60 per cent of the land, a ratio that changed little between the time of the first survey in 1928 and Hüsken's study in 1976–77. The main difference, by 1977, was that the percentage of land under sharecropping had increased, and more of the costs and risks of production had been transferred to sharecroppers, for lower returns.

The productivity increases of the green revolution were grafted onto a rural scene already characterized by "rent capitalism" and high rates of landlessness. As critics of the shared poverty literature emphasized, these were not new features of a rural economy "penetrated" by capitalism for the first time. Hence, it was no surprise that agrarian differentiation intensified as people with better access to land and capital, augmented by subsidized credit and farm inputs, took advantage of the new production regime. Many marginal farmers were squeezed out, but some also benefited from state investments in irrigation and infrastructure *if* they were able to hold on to their land. When intensively cultivated, even 1/12 of a hectare of good, irrigated land could feed a family for a year.[17] Yet the cost of inputs and credit presented a formidable risk of entering a downward spiral of debt. Far from being a level playing field, the "market" for credit was highly uneven: small landholders paid monthly interest rates 40–150 per cent higher than large landowners, who were able to use their assets and networks to access formal credit.[18]

A comparative study of three neighbouring villages on the north coast of West Java by Jonathan Pincus in 1989–90 provides nuanced insights into the changing relations between landed and labouring classes during the green revolution period. The weight of his findings, like those we have just reviewed, refutes Geertz's argument that poverty (or, more specifically, access to sources of livelihood) was generally shared among co-villagers acting in terms of what Geertz envisaged as the inclusionary ethic of "Javanese culture". Yet cultural precepts, or what we have called more broadly powers of legitimation, shaped the material outcomes that emerged. Some of the idioms Geertz reported—an obligation to favour kin and co-villagers in allocating access to land and work—came into play not as scripts for landowner conduct, but as grounds for mobilization among differently situated actors engaged in a struggle over which class would capture the gains from the green revolution.[19] Discursive legitimations were differently deployed, and differently effective, in each village. Village history, the degree of intimacy among co-villagers, and the capacity of landowners and workers to act collectively were crucial to explaining these variations.

In North Subang, first settled around 1910, repeated crop failures in the early 1960s forced many people to sell off their land at very low prices and undertake migrant farm work. Households with surplus rice bought up the land, leading to highly concentrated ownership. Soon after, a new set of migrant farmworkers arrived to work for these landlords. At the time of Pincus' survey in 1990, 4 per cent of households owned 73 per cent of the rice fields, 74 per cent of households owned no land at all, and 60 per cent had no access to land for rice cultivation.[20] Most labour was organized on an impersonal daily wage basis, and harvests were open to all comers regardless of village affiliation. The open harvest arrangement reflected the strong position of landowners vis-à-vis labour in North Subang. Ironically, it was also defended by North Subang workers, because their income depended upon open access to harvests in other villages. In the 1970s, when North Subang landowners attempted to close the harvest, North Subang workers mobilized successfully to forestall this plan because they feared that it would limit their access to wage work in other villages, where workers would exclude them in retaliation.[21] In this village, the landowners captured most of the productivity gains from the new technologies, and a seven-fold increase in profits enabled many of them to make the pilgrimage to Mecca.[22]

In East Subang, also settled around 1910, landownership was slightly less unequal. The top 4 per cent of households owned 45 per cent of the land; 50 per cent owned no rice fields. There was no influx of migrants from outside, a stronger web of social relations between households resulted in more tenancy and sharecropping, and there were fewer households (36 per cent) with no access to rice land for cultivation.[23] With the green revolution, both larger and smaller landowners shifted from open harvest to tied labour arrangements, in which a select group of co-villagers both planted and harvested the rice in return for specific shares of the crop. Their motivations were different: for the larger landowners, tying labour enabled them to recover more of the loans they had made to their sharecroppers and other debtors during the year, and to direct opportunities for harvest work towards their regular wage-workers as a means of labour discipline. For smaller landowners, labour tying enabled them to direct more work opportunities towards neighbours and kin, with the

　　　　　　　　　　　　　　　　　　　　　　Land Dilemmas in Southeast Asia

expectation that they would reciprocate. Dense social relations notwithstanding, these shifts in labour regime did not redistribute resources in favour of the poor. On the contrary, both small and large landholders reduced the shares they paid to their harvest workers as productivity increased, and took unilateral decisions to displace labour with new technology.[24] During a minor crisis in the form of a pest infestation, some landowners allowed their sharecroppers to postpone payment, but the consumption credit they extended to small farmers and labourers was at the standard interest rate of 50 per cent payable at the next harvest. Small farmers liquidated assets such as livestock, and sought off-farm work. One was forced to sell land, and four took on mortgages they had little prospect of repaying. More small farmers would have been forced to sell or mortgage land if the crisis had persisted for another season.[25]

South Subang was settled earlier, around 1890, and most of the current residents descend from the first settlers. The land was first irrigated in 1928, and the increased productivity enabled parents to partition land for their offspring. The result was a large number of micro-holdings, but still significant inequality. In 1990 the top 6 per cent of households owned 38 per cent of the rice land, 37 per cent of households owned no rice land, and 31 per cent had no access to land for cultivation.[26] The larger landowners' accumulation strategy included limiting the subdivision of their land and investing in education leading to salaried jobs. They were also involved in moneylending. The standard rate of interest on intra-village loans was pegged lower than in North Subang and East Subang, but it was still a hefty 30 per cent even among kin.[27] Farmers recruited most of their non-household labour force from within the village on a tied basis, an arrangement strongly defended by the large population of smallholders because it enabled them to complete field preparation tasks without a cash outlay. It also gave them access to portions of their neighbours' harvests, thereby spreading risk. Larger farmers complained of labour shortage but cited a "spirit of mutual help" within the village that prevented them from hiring outside labour. Workers also mobilized successfully to prevent landowners from adopting labour-saving technology such as herbicide. The material outcome of this resistance was that landowners used double the labour per hectare employed in neighbouring villages. Harvest shares had remained constant since the 1960s, and wage-workers successfully defended their relative share of productivity gains.[28]

Comparing across the three villages, there is no support for the thesis of "shared poverty" as a generalized village ethic, but the degree of intimacy among villagers—how closely they were related by kinship, and to what extent they shared a common history—stands out as an important variable in access to land and rewards to labour. The ability of small landowners to hold on to their land was affected by the agricultural wage, the availability of credit, the interest charged on loans, and the likelihood of debt forgiveness in the event of crop failure or illness, all of which were negotiated on a dyadic basis and reflected the particularistic relations between the parties. Even among close kin, however, a market calculus of short- or long-term gains was implicit in the positions that different actors took on the question of whether or not to extend access to land, work or credit. South Subang had the tightest kin links and an explicit discourse of solidarity that operated in tandem with worker mobilization to defend labour's share of the surplus product. The most important discursive effect of the notion of "the village" was not internal but in the assertion of a "moral boundary" in struggles over the use of outside labour in a context of labour surplus. This resonates with the situation in Cambodia we described in Chapter 3, where "community" is more important as a discursive means of negotiating resource claims with outsiders rather than as an organizing principle of internal relations.

A further restudy of North Subang in 1998 by Jan Breman and Gunawan Wiradi illustrates how the landed and landless responded to the entrenched inequality in this village. In the decade since Pincus' study, the landowners did not invest, innovate or diversify locally, although they continued to accumulate through land leasing and moneylending and used their privileged access to capital and village office to invest outside the village.[29] The landless continued to work in the agricultural sector in North Subang and across the coastal plain, where they roamed in search of daily wages and sometimes encountered violence from "local" wage-workers protecting their own niche. Remarkably, in a period when "deagrarianization" was assumed to be widespread, only 13 per cent of North Subang's economically active population was able to access rural off-farm work, and just 10 per cent were able to find work in Jakarta despite its being a mere four hours away by bus. Established Jakarta residents, organized territorially along ethnic lines, attempted to exclude them from both the formal and the informal sectors, and monopolized particular economic niches and living spaces. The only jobs they found were as unskilled labour in construction.[30] They were quickly expelled during the economic crisis of 1997, but—unsurprisingly—they were unable to find a foothold in the rural economy, which had already ejected them. Village landowners took no responsibility for their plight, and workers competed fiercely for the little work that was available.[31] The only positive shift during the period 1990–98 was the opening up of opportunities for women to work abroad as maids, in countries such as Malaysia and Saudi Arabia. In 1998, 57 women (from both landed and landless families) were abroad, the returns for their labour varying according to the debts they had incurred to secure their exit.[32]

Overall, research in Java confirms the continued relevance of land in both livelihood and accumulation strategies. Even when a significant proportion of total household income derived from off-farm sources (up to two-thirds in the nine villages studied by White and Wiradi in 1981), landownership continued to be highly correlated with returns to labour and capacity to invest (see also Chapter 5). In White and Wiradi's study, landless people, with minimal bargaining power, received the lowest pay, and their tiny capital limited them to the least remunerative petty trades. Larger landholders used the capital generated by their land (and subsidized state-supplied credit) to make loans, or to invest in improved technology, means of transportation and off-farm ventures. For modest landowners, off-farm work played an important role in enabling them to consolidate their position and avoid distress sales of land.[33] We return to the continuing significance of land as a basis of livelihood, and as the grounds of territorial belonging, in our discussion of mobilization in the next chapter.

Enclosure and Class Formation in Upland Sulawesi

In contrast to what Geertz envisaged—wrongly—as the stagnant egalitarianism of village Java, he described Indonesia's "outer islands" as a sea of (relatively) unchanging forest and swidden, punctuated by islands of dynamism in which agricultural innovation rapidly transformed landscapes, individualism and accumulation were rampant, and there was "great class differentiation and conflict".[34] This assessment, which he based on the colonial sociologist Schrieke's study of the coffee boom of 1908–12 in Sumatra, was largely correct. Throughout the colonial period, rapid uptake of cash crops among smallholders furnished important export revenues. It was also a cause for alarm among colonial officials, especially when farmers gave up planting food in favour of tree crops that tied up land for long periods and made them vulnerable to fluctuations in global market price (see Chapter 4). Nevertheless, wherever a nexus of market demand, sufficient infrastructure for market access, and rudimentary knowledge of a new crop's production requirements came together, farmers were quick to respond. Merchants played a key role in these transitions, supplying credit and often—though not always—ending up in control of land.[35]

Although the introduction of tree crops into swidden systems can have profound implications for land access, this is not always so. In the case of Kalimantan, described by Michael Dove on the basis of his research on Dayak longhouses in the 1970s, swidden and rubber production were complementary.[36] There was complementarity in the use of land and labour: rubber could be planted on land unsuited to rice, and farmers could postpone tapping during periods of peak labour on the rice swiddens. The mix also worked well because rice provided food and rubber provided cash. Since rice was often short, as swidden production was prone to failure, an income from rubber helped make up for the shortfall. Members of households with rubber trees could avoid wage work. In particular, they could avoid having to work for neighbours at critical times in the production cycle when they needed to concentrate on their own farms to secure the next harvest. They could also avoid having to borrow rice.[37] Rubber planting required no capital, hence no establishment debt. Future harvests could be pledged to a trader to gain access to credit when needed. Dove makes no mention of trees being seized by traders for non-payment. Most likely, the traders had no interest in managing rubber fields in remote upriver locations and had no need to do so if debt already bound the producers to them and yielded a handsome profit. Presumably, if their rubber trees were seized, farmers could plant more. In sum, traders profited, but planting rubber under these conditions did not set up a process of agrarian differentiation *among intimates*. If anything, it helped to forestall this process. Dove reports that households used tree planting as a tactic to exclude neighbours from a choice patch of land they wanted to hold, but such exclusion was not transformative. One person's rubber planting ventures did not exclude kin and neighbours from enjoying the same opportunity, because land was abundant and accessible to all. In this case, then, the market for rubber provided an opportunity, but there was no mechanism turning this opportunity into compulsion, and class differentiation did not occur.

The Kalimantan case described by Dove contrasts in every way with the case of cacao in upland Sulawesi traced by Li, in which upland farmers moved rapidly from a situation in which every farmer had access to land in 1990, to one in which, 20 years later, some highlanders had no access at all.[38] The process of exclusion was very intimate. Unlike the situation Li has described in another part of Sulawesi where migrants arrived to buy land for cacao,[39] in this case exclusion was a process initiated by the highlanders themselves, and the people they excluded were their kin and neighbours. Yet the highlanders had good reasons for wanting to switch to cacao. They had long been familiar with cash crop production, having grown tobacco since around 1820 and shallots from the 1950s, along with rice and corn for home consumption and trade. But they were chronically indebted to their tobacco traders and received low returns; their shallot production collapsed in the early 1990s due to fungal disease; and they experienced periodic famines when their swiddens dried up due to the severe droughts associated with the El Niño cycle. With cacao, they reasoned, they would have a source of income that would be more reliable than shallots, and they could still pluck the cacao when they were too old for the gruelling labour of swidden production. Back in 1990, they expected to maintain their food swiddens; but this plan changed as land became scarce and it made more sense to use it for the higher value crop. By 2009, the hillsides were covered in monocropped cacao. Indeed, smallholder cacao boomed across Sulawesi in this period, and Indonesia became a major world producer.

The first stage of exclusion involved farmers planting cacao on their current swiddens among the corn, effectively breaking the cycle in which a swidden would return after one or two seasons to the pool of land a group of kin inherited from the pioneer who had first cleared it. Most swidden systems in Southeast Asia recognize the labour of clearing primary forest as the basis for an exclusive right: the land pioneer can use the land, transfer it to someone else, or pass it on to his descendants. In many systems, the fallowed swidden land inherited from ancestors is divided among the heirs; but in this part of Sulawesi, it was retained in a common pool that any of the descendants was entitled to use and could also be freely lent to neighbours. Because the highlanders regard land planted with trees as permanent, private, individual property, planting trees was not just a change in land use, it was a complete change in the land's status. There was no talk of the land reverting to the collectively inherited pool after the tree cycle was complete. Nor were there customary rules forbidding the planting of trees. Rather, the individual enclosure of land in order to plant cacao was widely regarded as legitimate, and everyone joined in the process. The main rationale was a discourse of improvement. Highlanders argued that generations of swidden cultivation had left them impoverished, with none of the "progress" that had occurred on the coast (better houses, clothes, schooling). Everyone pinned their hopes on the new wealth that would be generated by cacao.

With a legitimating discourse in place, the main struggles that occurred in the initial stages of enclosure concerned who was entitled to plant trees on which patch of the collectively inherited swidden land. Individuals raced to plant cacao and lay claim to parts of this land pool before their co-inheritors did the same. A few kin groups settled the division by means of discussion, but for the most part it was the physical act of planting that excluded other potential claimants. In some parts of the highlands that had first been cleared generations ago, information about which ancestor had cleared which patch of forest was sketchy, and many highlanders were sceptical about the claims to descent that people who quickly planted large areas of cacao used to legitimate their actions. Once the cacao was in the ground, however, challenging the right of one person to exclude another from the patch of land in question meant disrupting valued social ties among kin and neighbours, and many highlanders were unwilling to do this. Ironically, strong social bonds among intimates *facilitated* their uneven exclusion. In a few cases, disgruntlement was expressed by force, as seedlings in disputed fields were mysteriously uprooted or burned at night. But even in these cases, the public narrative was that someone had accidentally set a fire by careless disposal of a cigarette, preserving at least the facade of harmony among kin.

People who fared well at the enclosure stage deployed a range of powers, among which the market was quite important. Those who had reaped profits from shallot production were able to hire labour to plant cacao quickly, while men and teenage boys from poorer families were often away seeking wage work outside the hills and missed the chance. Interventions by the Agriculture Department in the name of development had some uneven effects, as a few of the more ambitious farmers gained access to free cacao seedlings, enabling them to enclose more land. In addition to enclosing their inherited swidden land, highlanders also raced to enclose the remaining pockets of uncleared forest that served as a commons for hunting and for gathering rattan and timber. No one disputed the right of individuals to enclose this land, which fell under the customary right of a pioneer to clear new land to farm.

Stage two, the accumulation of advantages and disadvantages, set in very soon after the initial enclosure was complete and served to entrench the nascent inequality in land access. Farmers who had been successful in laying claim to more land were able to maintain their food production while waiting for their cacao to mature. In contrast, farmers who had claimed less land were unable to make ends meet during the transition period. Faced with the "simple reproduction squeeze", they began to treat their patches of enclosed land as commodities that they could sell to raise cash. Many farmers sold land with immature cacao to meet food needs during the El Niño droughts in 1993 and 1997, events that gave kin and neighbours with cash in hand the opportunity to accumulate. Gambling also played a role. Gambling had long been habitual among tobacco and shallot farmers, who often gambled away the proceeds from a season's crop; but so long as their land remained part of the co-inherited swidden pool, it was not theirs to stake. This changed after enclosure. The implication of indebtedness also changed. In the past, the only way a highlander could raise credit was by mortgaging a crop, typically tobacco, or by mortgaging labour by taking an advance for work to be performed in the future. Crop and tree mortgage systems quickly emerged for cacao, but if a debt became too large an individual landowner could be pressured to sell up. All the hillside farmers who successfully established themselves on a trajectory of accumulation supplemented their cacao incomes with moneylending. The legitimating discourse was one of helping: if they had money to hand, they could not refuse it to a sibling or cousin or other close kin who came asking for cash to buy food or pay for a child's wedding. Highlanders also borrowed money from non-intimates: the cacao traders who were willing to advance credit to secure a monopoly on the product. These traders monitored their debtors carefully to make sure they did not evade repayment by selling elsewhere.

Stage three, absolute landlessness, had emerged in the hills by 2006. Some households had accumulated up to 15 plots of land, purchased from kin and neighbours when they urgently needed cash. The capital for these purchases derived partly from their own farming and moneylending activities, and partly from the cacao traders, who had grown tired of supplying loans to farmers on a downward trajectory. The traders had no interest in claiming fields in default, as it was not practical for them to manage and monitor cacao stands scattered in the hills. Instead, the traders started to advance investment capital to the larger landowners to speed up the process of land concentration and consolidate their position as reliable farmers who could be counted on to repay their loans and sell cacao by the ton. Highlanders who lost their land worked as porters, carrying down the crop; but cacao farming absorbs little labour, and most of the work was done by the owners themselves, aided by chemical inputs such as herbicides.

By the time of Li's last visit in 2009, no one had been evicted, and very few of the landless families had left the hills. Farmers without cacao were still trying desperately to establish a few seedlings on tiny remnant patches of land or on very steep slopes, causing some rather spectacular landslides in the process. Lacking wage opportunities in the hills, the men were often away seeking work on the coast or in the headwaters where they hauled rattan and timber for minimal returns. Their wives and children continued to live on the land they used to own, serving as guards for the new owners and watching out for fires or theft—risks that were increasingly common as the major landowners moved their families to new houses on the coast, and evidence, perhaps, of the kind of covert protest action James Scott calls the "weapons of the weak".[40]

Contrary discourses emerged among highlanders as they contemplated the causes and consequences of exclusion from the land. The new landowners explained their accumulation in terms of their diligence and skill, and explained the condition of the poor in terms of personal qualities such as laziness, apathy and failure to plan. They suggested that there was still land available but the poor were too impatient to wait for crops to yield, preferring wage work with its immediate return. The poor, for their part, made a distinction between landowners who did indeed work hard, and those who only worked "with money" by employing others to sweat in their fields. They described their own situation in class terms: they lacked access to land, and they could not find work that paid a living wage. By 2009, the market price of land in the hills was so high they could never hope to buy any land; and even if they could access land, their daily struggle for survival made it impossible to invest time and money in cacao. Practically, the landowners were spared from frequent, painful encounters with impoverished neighbours and kin because the two groups that had come to occupy distinct class positions (defined by their ownership or non-ownership of the means of production) became less intimate: they withdrew from social contact across the class divide and associated mainly with kin and neighbours who shared their situation. The divide was consolidated by marriage strategies and patterns of investment, as the children of landowners attended school on the coast while the children of the poor remained in the hills—isolated, illiterate and unable to speak Indonesian or move about with confidence in the wider society. They desperately needed work, but there was little demand for their labour either on- or off-farm. Nor could they move off to a land frontier, because much of the potentially arable land in the district had been settled by migrants seeking land for cacao, enclosed for massive oil palm plantations, or designated for transmigration schemes to supply land to people from Java on the assumption that "the locals" had land to spare. All in all, their prospects were bleak.

Reflecting on how the four powers worked in this case, it might seem surprising that market power became such a key vector of exclusion among intimates. The highlanders had a particular reason for this. From their perspective, the virtue of a market transaction was precisely the way it cut through social relations. They found inherited land troublesome, as it was laced through with kinship and history, but land they had paid for was clearly their own. For this reason, many highlanders insisted on paying a sum of money to their siblings and cousins to legitimate their individual enclosure of co-inherited land, even when the enclosure was not contested. Highlanders who had sold their land to kin often regretted the sale—wishing they had not been so cash-constrained—but they did not argue that the kinship relation made the sale illegitimate, or perhaps reversible. They had sold the land, and that was the end of the matter.

Force in the form of arson played a role in disputes over particular plots of land, while theft came in as a protest against the condition of exclusion—but in neither case did force change the main outcomes. Custom regulated many aspects of the enclosure and accumulation process, most obviously through the notion that land became private property if a person had invested labour and capital in it. There was no custom to limit tree planting, or to ensure that enclosed land was evenly distributed. Old customs of reciprocal labour exchange and harvest shares were not rejected; they simply became irrelevant when rice and corn were no longer planted. State-based regulation and law were markedly absent. None of the land in the hills was titled, and most of the land sales were by verbal agreement among highlanders, usually without witness by a village official. Agriculture officials encouraged cacao planting as a way to increase production and bring the highlanders out of poverty—plans that concurred with the highlanders' own aspirations. However, government support was minimal, so for the most part the new land regime in the hills was the result of the highlanders' own initiative. When they began the process of exclusion, the highlanders imagined it would be mutual: everyone would enclose land, and they would all prosper through cacao. They did not anticipate the unequal outcome of the initial enclosure process, or the process of agrarian differentiation that would follow. Their hope was to improve their economic condition, a goal that was broadly shared though unevenly accomplished.

New Exclusions in Village Vietnam

In this final section, we examine the powers at work in the re-emergence of landlessness in village Vietnam in the period since family farming was formally relegitimized in 1988. Struggles for access to land in Vietnam were central to the movement for independence—as important, for many rural people, as the removal of the French. The Viet Minh seized property from fleeing landlords and distributed it more or less equally to the landless. An estimated 73 per cent of the northern population benefited from land distribution by 1956.[41] There followed a period of collectivized agriculture (extended incompletely to the south after 1975), but returns were very low, leading farmers to quietly expand household production. Under the new rules stipulated in 1988, families were granted use rights to land for a period of 15 years for annual production, and 40 years in the case of perennials.[42] The 1993 Land Law that superceded this decree allowed for individual landholdings under 20-year leases with a 3 hectare ceiling for annual cropland, and 50-year leases with a 10 hectare ceiling in the lowlands and a 30 hectare ceiling in the uplands for land planted to perennials.

Researchers concur that the post-collective reallocation of farmland to families after 1988 was relatively egalitarian. However, farmers demanded the right to pass land to their offspring, lease it, mortgage it and "transfer" it. The result was a de facto land market, ushering in the possibility that land could be accumulated and some farmers excluded by the "everyday" processes we have already explored: harvest failure, illness, debt, uneven access to credit, variable efficiency related to scale, and the squeezing out of producers who lacked the resources to compete.[43] Anticipating this eventuality, signs of which had already begun to emerge in 1991, Haroon Akram-Lodhi and other researchers began to track changing patterns of access to land, and to explore the mechanisms that foster or mitigate the re-emergence of agrarian classes. The overall pattern revealed in the macro data is one of increasing inequality in land access. Landlessness increased from 8.2 per cent in 1993 to 18.9 per cent in 2002.[44] By the late 1990s there were once again private landowners—often party officials—who owned estates well in excess of the legal maximum (5 hectares, and in some cases as much as 1,000 hectares).[45] Yet the patterns were spatially uneven as villagers accepted exclusion among intimates as inevitable in some cases, but worked against it in others.

In the highlands of the northwest, studies by Thomas Sikor highlight the persistence of popular support for a system of land distribution in which land access is linked to family size. In a minority Black Thai village, where the customary practice of periodic land redistribution based on family labour capacity preceded socialist rule, villagers rejected the government plan to allocate land use rights to families for a 20-year period.[46] They insisted on retaining collective, village-level control over their main rice fields, including the right to reallocate fields periodically. Some villages took the collective decision to extend the principle of egalitarian redistribution to upland fields as well, especially as those fields became more valuable and scarce. When the Forest Department allocated forestland to particular households and charged them with maintaining it under forest cover, villagers again rejected this attempt to regulate their land relations. They continued to expand agriculture onto forestland, and to allocate use rights flexibly among themselves. On all these points, the authorities were obliged to accommodate villager practices by adjusting the rules or abandoning attempts to enforce them.[47] As a result of these practices, economic differentiation within the Black Thai villages Sikor studied was limited and non-cumulative, reflecting mainly differential access to family labour as new households were formed and matured. There was no wage labour. The only households excluded from access to land were those that failed to meet their obligation to participate in labour service to maintain village facilities, or amassed debts to the collective they could not repay.

Remarkably, a neighbouring village studied by Sikor and Thi Tuong Vi Pham that was populated by lowland Kinh who had volunteered for resettlement to the uplands in 1963 also elected to maintain collective control over their rice land, and the right to redistribute it periodically. They carried out reallocations in 1985, 1991 and 1998, based on household size. They had their own rationale for taking this approach. They argued that they had endured the hardships of resettlement and developed the land together, so they should continue to access it in an equitable manner. Unlike the Black Thai, they excluded upland fields from the reallocation, and also excluded the patches of land some households had developed for intensive vegetable production.[48] The outcome of collective control over rice land in this village was somewhat different from the Black Thai examples. Since their capacity to expand their agricultural activities was limited by the lack of a forest frontier, they began to invest in trade, transport, processing and manufacture, taking advantage of their command of the Vietnamese language, lowland contacts, and strategic location beside a major road.[49] Sikor and Pham argue that the Kinh villagers' energetic diversification and entrepreneurial risk taking were enabled by the fact that every household had secure access to a share of the village rice land as a non-alienable, subsistence "floor". In this village, as in the Black Thai villages, wage labour relations did not develop, presumably another feature of the villagers' access to land for rice: they were not compelled to do wage work as a condition of survival. There was economic differentiation, as some households were more successful than others in building up capital to expand their enterprises, but there was no separation into landed and landless classes. Further, the diversified economy yielded new opportunities, some of which required little or no entry capital.[50] Rather than essentialize the difference between Kinh and Black Thai, Sikor and Pham point to the contingency of these outcomes, as the hardships of resettlement and a limited land base turned unexpectedly into assets for this particular group of Kinh at this time and place. They also point to the role of agency and the power of legitimation. The Black Thai reshaped official land policy and practice according to their view of what was right and proper, while the Kinh both created new markets and designed their own system of land management to limit the inequality that emerges when markets dominate.

A contrasting case study in the northern uplands by Bent Jørgensen describes how district officials colluded with village-level officials to manipulate Land and Forest Allocation programmes to keep minority people (Dao, Hmong and Muong) from gaining access.[51] An important power of exclusion in this case was the legitimation supplied by the discourse of efficiency and "development" juxtaposed—but not reconciled—with the discourse of poverty alleviation. Village-level officials argued that the degraded forestland the Forest Department was allocating to families for replanting, together with subsidized credit and farm inputs, should be allocated only to progressive farmers who could make the best use of them, and not to the poor—whom they characterized as lazy, stupid and undeserving.[52] Although the local officials (who doubled as village landowners and moneylenders) were implicated in the impoverishment of the people they branded undeserving through "everyday" practices such as charging 100 per cent interest on pre-harvest loans, they were reluctant to admit their role because these practices violated a sense of "right conduct" among co-villagers that still had moral weight.[53] They described their moneylending as "helping". They also confused their official and private roles by encouraging poor farmers to borrow from formal credit schemes, knowing that when the farmers could not repay they would be obliged to mortgage their land to a moneylender—often the same village official who had arranged the original loan. To disguise their land-grabbing under the forest allocation scheme, local officials deflected responsibility upwards, arguing that land allocation was handled by their superiors, so they were not responsible for the uneven results.[54] Some grabbing of "development" resources was blatant. One village official responsible for distributing inputs to co-villagers allocated himself 400 kilograms of subsidized fertilizer, 25 tree seedlings, a pig and two chickens, plus a significant loan of cash.[55]

In this case, as in that of green revolution Indonesia, the uneven distribution of "development" handouts intersected with processes of agrarian differentiation that were already proceeding on an intimate scale. Selected villagers with access to land, capital or the prerogatives of office were enabled to further strengthen their position vis-à-vis the poor. This process was not new. During the collectivization era village officials and party cadres were often able to acquire more than their share of prime resources, and the advantages they gained in that period were carried over into the post-collectivization period when superior access to land, labour, information and capital enabled them to accumulate.[56] In Jørgensen's field sites, in contrast to the cases described by Sikor and Pham, there was no counter-discourse promoting allocation based on need.[57] Instead, Jørgensen argues, the persistent authoritarian features of the regime combined with idioms of development to produce a double jeopardy for the poor.[58] This did not mean, however, that the rising inequality and its causes were unobserved. Although they did not use the language of class differentiation, since class is not supposed to exist in socialist Vietnam,[59] officials Jørgensen interviewed quickly classified villagers into rich, middle and poor, and readily explained how everyday processes of accumulation were operating in their midst. The chairperson of the People's Committee supplied Jørgensen with an astute analysis that highlighted elite land-grabbing and commented on how rich villagers and local officials used a discourse of laziness to justify their unjust practices.[60] From this we conclude that the discursive legitimation of exclusion in terms of laziness was not especially effective, since its flaws were too obvious. The main power village officials and landowners used to exclude co-villagers from access to land and development resources was the power of regulation, suitably manipulated, combined with the power of the market in the form of cash to extend loans and entrap people in exploitative relations.

In the same period that agrarian classes distinguished by access to the means of production were emerging in some upland areas, they were re-emerging in the lowland deltas as family-based farming was re-established. This outcome was anticipated in an anonymous article (quoted by Jørgensen) published in a Vietnamese journal in 1964, which reminded readers how quickly class differentiation re-emerged in the Red River Delta in the 1950s, after the first land reforms that distributed land to landless households, as farmers facing drought, illness or other difficulties fell into debt.[61] The Viet Minh opted for collectivization for the express reason that it was the only means to prevent the re-emergence of landlessness by eliminating all opportunities for usury and mortgage. Yet millions of peasant households resisted collectivization and succeeded in regaining the right to family-based farming. Further, like the households in Jørgensen's study, they wanted their own land, despite the risk that a spiral of debt might cause them to lose it.

The Mekong Delta had the most pronounced degree of landlessness by 2000, with estimates ranging from 5 per cent to 30 per cent.[62] The delta is highly productive, capable of producing multiple crops of rice as well as fruit, vegetables, and farmed fish and shrimp. The result is high incomes for farmers with the capital to invest in high-tech, high-risk ventures, and an acute spiral of dispossession for others. According to a study by Philip Taylor, the principal vectors of "everyday" agrarian differentiation in the delta around 2000 were uneven access to non-land capital (including overseas remittances) and to information that yielded material advantage. For example, advance knowledge of where a new road would be built enabled local officials and their allies to buy up land cheaply from impoverished farmers. For villagers with capital, the most secure pathway to accumulation was not farming; it was the supply of agricultural inputs, processing, transport, services and credit, all of them beyond the reach of the poor. Further, the high degree of mechanization meant there was little demand for labour, and the landless were obliged to roam far and wide in search of work, losing the intimacy of attachment to a "home" village in the process. Children forced to follow their parents from place to place could also lose access to education, which put in jeopardy their capacity to belong to the nation as "normal" citizens.[63]

In the Red River Delta, in contrast to the Mekong, studies around 2000 reported persistent moral force and everyday support for the notion of a village as a social and territorial arena within which access to land should be broadly distributed. Benedict Kerkvliet reports that Kinh farmers in the rice heartland continued to support the idea of periodic readjustment of the rice land to maintain equality in access, taking into account issues such as family size, responsible use, local residence and a commitment to meet obligations to the collectivity. Absentees were to be excluded.[64] A village study by Nguyen Van Suu confirmed that households in the delta accepted the notion of state ownership and fully expected that the authorities would carry out a redistribution when their current use rights expired. However, they believed the adjustments would be minor, and they were investing in production with the expectation that they would retain control of most of their current land.[65]

Taken together, the studies from Vietnam that we have reviewed here offer striking insights into a possible resolution to the conundrum of exclusion at the intimate scale. This is a resolution that would work by tempering market processes, giving everyone just enough exclusionary power to enable them to farm productively without enabling them to accumulate land to the exclusion of their kin and neighbours. Sikor's studies of upland villages show one way this can be done, by allowing market processes full reign in some spheres of the economy, but *not* in relation to rice land. This was a measure that some Kinh and Black Thai villages insisted upon in opposition to the programme of district officials to distribute alienable rights to individual households. Another approach would remove responsibility for redistribution from the intimate sphere of the village, where rampant inequalities and a limited sense of community (as in the highland villages described by Jørgensen, or the Mekong as described by Taylor) make it unlikely that village landholders, who often double as village officials, would agree to redistribute. Instead, redistribution would need to be imposed by the central government, making use of the provision in Vietnam's land law that reserves the state's right to regulate and reallocate land periodically, and grants landholders only time-limited use rights, not outright ownership.[66] This approach has popular support in the Red River Delta, as Kerkvliet and Nguyen Van Suu's studies show. It has variable support at the state centre, where some senior officials consider land concentration necessary to increase productivity and economic growth, while others worry over increased landlessness and its potentially destabilizing effects.[67]

Kerkvliet, a close observer of changing agrarian relations in Vietnam, retains his faith in the "everyday" power of the Vietnamese people to temper market processes by reference to family, community and socialist values, just as they tempered the excesses of collectivization by quietly reasserting family-based farming practices even while they remained illegal.[68] It remains to be seen which set of social forces will predominate when the current set of land use rights begins to expire: the forces that will mobilize to defend villagers' exclusive right to land they have seized or accumulated through market means, or the forces that will mobilize to insist upon redistribution.

Conclusion

In this chapter we examined how powers of exclusion operate at intimate scales, among kin and co-villagers who are launched on different trajectories, some accumulating land and capital, while others are progressively dispossessed in a process of agrarian class formation. The main finding was that market powers and legitimation are closely entwined. It takes powers of legitimation to insert market principles in transactions between intimates, and the attempted legitimation may be contested. Struggles over resources are, simultaneously, struggles over meaning, in which the outcomes are far from guaranteed.

Local agency matters in resource struggles, but the conditions are not made by villagers alone. They are the cumulative outcome of processes, relations and deployment of powers that span different scales. The discourse of development that bets on the strong in order to secure economic growth has supplied an important mode of legitimation for processes of class formation at the village level. Yet this idiom is not only purveyed from the top down, it intersects with locally generated concepts of diligence and sloth that distinguish the deserving from the undeserving on the basis of their personal qualities. Similarly, good prices for tree crops on global markets can stimulate a boom, but such a boom will occur only if it intersects with villagers' capacities and plans. Market relations—like other relations—need to be produced and rendered legitimate. Once produced, however, they confront people as social facts that cannot be ignored.

We have shown that images of the harmonious Southeast Asian village, comprising homogenous, egalitarian peasant smallholders, were exaggerated, and that processes of exclusion and accumulation among intimates can emerge with surprising speed. Yet our findings indicate significant variation. It is important to understand why land enclosure through tree planting led to rapid class differentiation in Sulawesi but not in Kalimantan; why the landless villagers of South Subang were able to make some gains in the green revolution while those of North Subang could not; and why Vietnamese farmers might continue to debate the pros and cons of a land market. This type of analysis does not resolve the conundrum of exclusion, but it does help to expose the different ways the conundrum is understood and worked upon by socially situated actors. As we demonstrated in Chapter 2, processes of intimate exclusion bedevil programmes for land reform that aim to create a countryside in which every peasant household that has been allocated a plot of land will continue to possess it —no more, and no less—in a self-sustaining equilibrium. We return to this point in the next chapter, where popular mobilizations for access to land also routinely overlook intimate exclusions among villagers, to focus their energy on countering the powers of exclusion wielded on a large scale.

Notes

1. See Chapters 4 and 7 regarding the coffee boom in Vietnam. See Li 2007c on tensions between migrants and locals in the context of the cocoa boom in Sulawesi.
2. See White 1989: 19–20.
3. Brenner 1985, Wood 2002.
4. Bernstein 1994: 56.
5. We examine this debate in the context of Java, but there were parallel debates about overly romantic portrayals of village Thailand and village Vietnam, the latter pitting the arguments of Samuel Popkin 1979 against James Scott 1976.
6. Geertz 1963: 97.
7. We draw closely on the critical assessments of Geertz and related studies by White 1983, Alexander and Alexander 1982, Hüsken and White 1989, Hüsken 1989, and Alexander and Alexander 1991.
8. Numbers for landlessness in the early 19th century are sketchy, but see Elson 1997: 123. From surveys in the late colonial period, Alexander and Alexander 1982: 603 cite 38 per cent landless in 1926 and 50–60 per cent in 1965. The Declining Welfare Survey of 1904–5 found that around 40 per cent of the total population of Java was landless, and 5 per cent of the population owned more than 30 per cent of the land (Hüsken and White 1989: 240–1).
9. Alexander and Alexander 1991; Henley 2009: 7.
10. Hüsken (1989) tracked a village in north Central Java, where, in the 1920s, 40–50 per cent of villagers owned no land and some village officials owned up to 50 hectares. These officials were descendants of the pioneers who had settled the area in the 19th century. On frontier processes, see also Breman 1980, 2000; Boomgaard 1991; and Wolters 1999.
11. In precolonial Sulawesi, for example, about a third of the population was "in some form of slavery", mainly due to debt (Henley 2009: 4, 22). On the prevalence of debt slavery across the region, see Reid 1983.
12. Hüsken and White 1989: 240–2; Hüsken 1989: 318; Alexander and Alexander 1991; Li 2007c.
13. White 1983: 26–8; Alexander and Alexander 1982: 602–5; Alexander and Alexander 1991; Hüsken and White 1989: 244–5. Hüsken 1989 emphasizes cyclical commercialization and decommercialization.
14. Alexander and Alexander 1982: 601, 608, 610–1, 615.
15. White 1983: 19, 28–9; Hüsken and White 1989: 248.
16. Hüsken 1989. See also White and Wiradi 1989: 280–3.
17. Hüsken and White 1989: 252–8; Hart 1989; Hüsken 1989: 310–1.
18. Pincus 1996: 149–52.
19. For further accounts of the varied outcomes of the green revolution and the implications of shifting land and labour regimes, see White 2000; Alexander and Alexander 1982: 613–4; Hart 1986; and the studies in Hart, Turton and White 1989.
20. Pincus 1996: 22, 40–1, 204.
21. Pincus 1996: 102–8.

22. Pincus 1996: 135.
23. Pincus 1996: 24, 41, 206.
24. Pincus 1996: 109–17, 124. See also White and Wiradi 1989: 289.
25. Pincus 1996: 152, 176–9.
26. Pincus 1996: 24–5, 41, 205.
27. Pincus 1996: 152, 169, 171–2.
28. Pincus 1996: 117–23, 135.
29. Pincus 1996: 170; Breman and Wiradi 2002: 102–3, 140–3.
30. Breman and Wiradi 2002: 68–71, 111, 144.
31. Breman and Wiradi 2002: 154–5, 166, 307.
32. Breman and Wiradi 2002: 116–28.
33. White and Wiradi 1989: 291–9; White 1991: 61.
34. Geertz 1963: 120. "Class" and "conflict" are terms absent from his description of Java.
35. Elson 1997: 98–102, 131–3, 197, 210.
36. See Dove 1984, 1993. This complementarity is also explored by Cramb 2007a. Cramb 2011 describes the differential uptake of rubber by different social groups in Sarawak, and its impact on accumulation.
37. Geddes 1961 observes that loans of rice among Land Dayaks in Sarawak carried interest within the community of 50–100 per cent. Thanks to Rob Cramb for this observation.
38. This section is based on Tania Li's field research, reported in Li 1996, 2001b, 2002c, 2010a.
39. See Li 2007c and the comparison between the two sites in Li 2002c.
40. Scott 1985.
41. Kerkvliet 2006: 285–6; Nguyen Van Suu 2007: 311.
42. Kerkvliet 2006; Nguyen Van Suu 2007: 323.
43. Akram-Lodhi 2005: 74, 79, 85–6, 92–3.
44. Akram-Lodhi 2005: 84–5. See also Kerkvliet 2006: 296.
45. Akram-Lodhi 2005: 88.
46. Sikor 2001, 2004.
47. Sikor 2001: 944; Sikor 2004: 89–90.
48. Sikor and Pham 2005: 410, 417–8.
49. Sikor and Pham 2005: 412–7, 423.
50. Sikor and Pham 2005: 418–23.
51. Jørgensen 2006: 118–24, 149–51, 154. Jamieson, Cuc and Rambo (1998) provide an overview of the marginalization of minorities in the northern highlands.
52. Jørgensen 2006: 132–8, 144.
53. Jørgensen 2006: 222.
54. Jørgensen 2006: 188, 210, 214–9, 233.
55. Jørgensen 2006: 163.
56. See Tran Thi Thu Trang 2010 for a village-level study of patterns of accumulation and loss.
57. Jørgensen 2006: 117.
58. Jørgensen 2006: 235.

59. Jørgensen 2006: 208–14.
60. Jørgensen 2006: 226–7.
61. Jørgensen 2006: 165.
62. Kerkvliet 2006: 296.
63. Taylor 2004. See also Marsh and MacAulay 2006.
64. Kerkvliet 2006: 294–5.
65. Nguyen Van Suu 2007: 319, 327.
66. Kerkvliet 2006: 289, 300. Parsons 1974 discusses the risks of "liberty-exposure" when land becomes a commodity, and the need to devise market-oriented arrangements that stop short of full alienability. Li 2010b offers an extended discussion of colonial and contemporary attempts to achieve this goal.
67. Marsh and MacAulay 2006: 10–3.
68. Kerkvliet 2006: 300.

7

Counter-exclusions: Collective Mobilizations for Land and Territory

The final process we examine is collective mobilization by groups of people seeking to counter their exclusion from land as territory or productive resource, and to assert their own powers to exclude. Legitimation plays a central role in collective mobilizations, as people draw allies to their cause and express their claims in terms such as the right to livelihood, social justice or territorial belonging. Collective mobilizations also contest the discourses that legitimate state-driven regulatory exclusions (for dams, conservation areas, plantations) in the name of development and the public good, but fail to deliver compensation or other benefits to the people who lose their land. Force enters into collective mobilizations as massed protesters are confronted by violent repression, or themselves engage in violence to reclaim land. Market powers play a more limited role. Although they shape exclusionary regimes at multiple scales, they do not offer an obvious target for counter-attack. In contrast, reclaiming land from state authorities, large corporations or people defined as outsiders who do not belong are programmes that can mobilize broad constituencies.

Although the deagrarianization thesis holds that land is becoming less and less important for livelihoods in the global South, collective struggles to defend, reclaim or gain access to land are widespread in the region. One indicator of their prevalence is the frequency with which they are reported. In Indonesia, for example, an inventory of land conflicts reported in the media between 1972 and 2001 documented 1,475 cases in 2,277 villages covering 2.5 million hectares and affecting almost 2 million people—no doubt an underestimate, since the reporting was biased towards Java and Sumatra, and through most of this period the media was far from free.[1] We can assume that individual or covert actions to defend or gain access to land were far more widespread. Our focus here, however, is on large-scale collective mobilizations involving coalitions or

170

networks of actors that have persisted over a period of months or years and have garnered significant public attention through practices that range from presenting petitions to organizing demonstrations, lobbying for legal and policy changes, blockading roads, occupying land, and finally, at the violent extreme, engaging in mass evictions.

Diverse as they are, two features of contemporary mobilizations stand out. First, the state is often prominent as the party against which claims to land are directed. Protests highlight harmful policies and broken promises. Yet they do not propose to take over the state apparatus, unlike the Communist-inspired movements of the 1950s and 1960s that sought to dislodge the feudal-bureaucratic class that exploited peasants and treated them with contempt. The prominence of state policies and corrupt practices as the target in struggles over land is a direct result of the strong role taken by ruling regimes of all ideological stripes in the allocation of land for purposes such as population resettlement, urban development, agro-industry, dams, logging and conservation. It also reflects the gap between the promises that national regimes in the post-independence period have made to their citizens—development, security, respect for basic rights—and what they have delivered. Across the region, patterns of mobilization indicate that this gap returns to haunt regimes that ignore the aspirations their own promises have stimulated. Even the most authoritarian regimes promise development and seek legitimacy in these terms. They can be challenged on internal grounds: were planning processes carried out fairly and transparently, were resources properly used, and were rules for land use, environmental assessment, compensation and so on followed?[2] Or, they can be challenged at the more fundamental level, when a counter-discourse questions the state's right to allocate land, or to define what counts as development.

The second feature is that identity defined not in class terms, but in terms of ethnicity, culture or attachment to a place, is prominent as the grounds on which groups mobilize to contest exclusions and assert claims. Yet, as we have stressed throughout this book, exclusion has a double edge: every counter to one discourse of exclusion necessarily proposes exclusion on other grounds. When different grounds for legitimating exclusion are contradictory—historical identification with a place versus need for land, for example—there are not just debates but on-the-ground struggles over who should be given priority, and who should decide.

To explore this vexed terrain, we have organized this chapter around four of the more prominent legitimating discourses that have animated popular mobilizations over land over the past two decades. First, we consider the cause of people who have mobilized under the banner of indigeneity to counter discourses that hold them responsible for environmental degradation, and argue for their right to access forested and hilly land on the grounds that they have a unique capacity to manage such land sustainably. Conversely, people who do not have this capacity should be excluded. We show how this strategy has played out very differently in contrasting cases, the Philippines and Thailand.

Second, we examine large-scale mobilizations to restore ethno-territorial autonomy in contexts where people defined as outsiders have become numerous as a result of spontaneous migration and government programmes. The legitimating discourse here is that historical ties to a place take precedence over claims based on arguments such as "We bought this land in good faith" or "The government put us here". Our illustrations are from Indonesia and Vietnam.

Third, we consider mobilizations animated by the notion that citizens have the right to land as a basis for livelihood. This discourse depends upon an affirmation of the state's right to allocate land through land reform, an affirmation that runs up against the right of ethno-territorial groups to determine who belongs on "their" land. We examine the tense encounter between divergent streams of mobilization in the case of Indonesia.

Finally, we examine how people have mobilized against evictions, when proponents argue that these evictions are necessary for national development and the common good. The counter-discourse, in these cases, tends to draw upon images of rural utopias to emphasize the attachments of people to place, and to wholesome ways of life. Our main example here is Thailand's Assembly of the Poor, clear enough about what it opposes—evictions and unfair treatment—though more starry-eyed than many rural Thais about the quality of life in the past.

"Indigenous People" as Environmental Managers

The question of who qualifies as indigenous in Southeast Asia is a complex and unresolved question.[3] Nevertheless, mobilizations by people using the banner of indigeneity to assert their right and capacity to manage forested and hilly lands have become common in the region. In significant part, the notion of indigeneity took shape as a counter-discourse intended to refute the claim that shifting cultivators were forest destroyers who put development at risk by causing floods, landslides, and the silting of rivers and dams needed for lowland irrigation. All over Southeast Asia, people described as highlanders or hill tribes have been defined as squatters on state land, harassed, evicted and forcibly resettled because of their supposedly delinquent practices. Their land has been zoned as political forest, deemed "empty", allocated to agribusiness, or designated for settlement by lowland peasants arriving spontaneously or through state-sponsored schemes. To counter the environmental discourse used by state agencies to legitimate programmes to exclude them, these groups have attempted to reverse the negative characterization of their land and resource use practices, and argue that it is state-licensed logging, settlement and agricultural expansion that destroy forest, while their customary traditions of resource stewardship guarantee sustainable management.

The most vigorous and effective campaign to secure rights to land on the basis of this discursive reversal occurred in the Philippines. As we explained in Chapter 2, the Indigenous Peoples' Rights Act was passed in 1997. This followed two decades of lobbying by a network that included members of what were then called "cultural communities" as well as legal scholars, foreign experts, donors, activists in the NGO movement, and former activists working within the state apparatus. Under the terms of the act, 6 million to 12 million people (out of a population of 89 million) could qualify as indigenous people, with the right to manage the ancestral domains they had occupied since time immemorial. We drew attention to two key features of the land rights conferred under the act: that land is to be held communally, and it is inalienable. Here we want to highlight a third feature: the requirement that people who claim the right to land through this act must manage their land sustainably. This requirement carries through the advocates' argument that "indigenous people" have unique capacities for environmental management, and it obliges them to perform as billed.[4]

Applications for recognition of Ancestral Domain must be accompanied by an Ancestral Domain Sustainable Development and Protection Plan that commits the applicants to the collective management of resources and maintenance of the ecological balance. These are onerous conditions that can be difficult to meet. In contexts where they are short of land, the people claiming an Ancestral Domain face the same pressures as other peasant cultivators who farm intensively on steep slopes and use chemical inputs in order to make ends meet in a competitive market system. The requirement for collective management presents a problem in conditions where individual land tenure, cash crop production and de facto land markets are rapidly emerging, or already well established. Taking responsibility for excluding non-members—neighbouring groups, and the tens of millions of lowland Filipinos who seek livelihoods in the uplands—is a heavy burden. Conflicts have erupted as groups excluded from land claimed as Ancestral Domain dispute the claimants' stewardship capacities, or shift the legitimating discourse, arguing that their superior farming skills enable them to contribute more to national development. Where Ancestral Domains overlap protected areas, as they often do, the management burdens are even more severe, and they demand detailed negotiations with officials, donors, NGOs and technical experts, each with their own priorities. Thus, the assertion that "indigenous people" are good resource managers has successfully reversed one of the grounds for their exclusion, only to institutionalize another, as people who claim an Ancestral Domain but fail to live in harmony with nature risk having their land rights challenged or abrogated. Their discursive counter to exclusion, in short, can backfire.[5]

In Thailand, in contrast to the Philippines, prospects are remote for the recognition of highlanders as "indigenous people" with distinct rights to land legitimated in terms of environmental stewardship. Some of the upland minorities—known in Thailand as hill tribes—are viewed as non-Thai and are wholly or partly excluded from the privileges of Thai citizenship, even when they have lived within the national border for decades, if not centuries.[6] There is some dispute over the number of hill tribe people in Thailand, but official figures suggest around 700,000 in a population of 63 million.[7] These people are not merely ethnically and culturally distinct. As Peter Vandergeest argues, their "non-Thai" designation is racialized, that is, treated as natural, essential and permanent. Authorities take their residence in hilly terrain, their transgression of political forest boundaries, and practice of shifting cultivation as indicators of the hill tribes' alien character. The fact that about 10 million ethnic Thais also live in the political forest, and that some of them practise upland shifting cultivation, is consistently overlooked in academic, policy and journalistic debate.[8]

Thailand's hill tribes have pursued different strategies in mobilizing over land. A few hill tribe groups, notably the Karen, have successfully reversed the forest-destroyer discourse to gain recognition as forest stewards.[9] For most groups, however, hill tribe status continues to be a liability that makes them vulnerable to eviction. Some Thais reserve special opprobrium for hill tribes such as the Hmong, many of whom have prospered through commercial agriculture. This opprobrium is based on the lack of fit between Hmong practices and characteristics that elicit a degree of sympathy, if not understanding: hill tribes must either be poor and backward or traditional and conservationist, but the Hmong are neither.[10] In 2000, a "deep green" NGO joined forces with the forest department to encourage valley farmers to burn down the fruit tree groves of Hmong farmers in the hills above the valley on the grounds that the Hmong were destroying the watershed, and that they were not Thai.[11]

Reacting against these racialized exclusions, advocates for community forestry seek to extend forest access to all uplanders on the grounds of their rights as citizens, rights they argue should be extended to hill tribes. Yet their arguments still draw elements from the counter-discourse of indigeneity, as they construe uplanders as capable forest stewards, and emphasize the non-commercial use of land and forest, with all the limits entailed.[12] In these cases, using powers of legitimation to reverse exclusion may work for one group of highlanders but have the unintended effect of intensifying the exclusion of others, when they cannot fit the stewardship niche carved out by the reversal strategy.

Reclaiming Ethno-territories

Around 2000, both Indonesia and Vietnam experienced large-scale mobilizations by ethnic minorities seeking to reclaim the territories in which they were historically dominant, and relatively autonomous, before the arrival of migrants in their midst. These struggles had some overlap with the claims of people who organized under the banner of indigeneity to counter their exclusion from the political forest, but we treat them separately because the scale was much larger, involving entire provinces or districts, and because capacities for environmental stewardship were not key to the legitimating discourse. Access to land as a productive resource was a flashpoint in both these mobilizations, but their main focus was on land as territory imbued with the history and identity of a particular group, and the argument that only members of the ethno-territorial group in question had the right to define who could access resources and wield exclusionary powers.

In Indonesia, with its famously diverse cultural mosaic, the number of groups that could potentially make claims of an ethno-territorial kind is large—hundreds if the scale is the district, thousands at the level of sub-district (roughly 20,000–50,000 people). In most districts and sub-districts, one ethnic group, usually marked by its use of a distinct language, is recognized as the original (*asli*) population. Some ethno-territorial groups retained a degree of autonomy under Dutch colonial provisions for indirect rule, although they lost ground when Suharto's New Order regime introduced a standardized, national framework for the administration of people, territory, and the vast land and forest resources to which the state laid claim. As the tight controls of the Suharto period were lifted, popular mobilizations demanding ethno-territorial autonomy re-emerged at multiple spatial scales. These mobilizations were encouraged, and to some extent provoked, by legislation to decentralize administrative authority to district level that went into effect in 2001. Decentralization raised the stakes of control over the district state apparatus, since districts were granted more control over revenues and resources. Democratization added to the tensions, since key district positions would henceforth be elected, and aspirant politicians began to mobilize constituencies along ethnic (or sometimes ethno-religious) lines. There was a rush of administrative fragmentation, as new districts were formed by actors seeking to bring ethnicities, territories, vote banks, resource revenues, and access to government jobs and spending streams into alignment.[13]

The decentralization legislation reinvigorated long-standing debates about who had the right to decide whether or not someone was entitled to acquire land, use resources, or take up residence when they arrived as "guests" in the ethno-territory of another group.[14] Even before the legislation went into effect, there were instances of violent eviction carried out by mobilized groups attempting to reassert their territorial sovereignty on a large scale. One such eviction occurred in Aceh in 1999 as part of the ongoing struggle between Acehnese separatists and the central regime. Transmigrants from Java who had been placed in Aceh under government programmes over a period of several decades, and whom many Acehnese saw as instruments of New Order opposition to Acehnese autonomy, received letters from the armed insurgents of the Gerakan Aceh Merdeka (Aceh Freedom Movement) ordering them to leave and threatening violence.[15] The Department of Social Affairs reported that 34,000 Javanese had fled Aceh by the middle of 1999.[16] Evicted transmigrants were housed temporarily in the compounds of the Transmigration Department in Aceh before they were transferred to Java, where they were crammed into the transit centres usually used for lodging transmigrants who were outward bound. These centres, moreover, were already crowded with ex-transmigrants from other islands also fleeing violent attack.[17] Rather than acknowledge these evictions as a symptom of revived ethno-territorial claims and a popular challenge to the state's right to decide who should live where, the Transmigration Department seemed determined to replicate the problem. It retooled itself as an agency for resettling refugees, including former transmigrants, to new transmigration sites in Sulawesi, Kalimantan and wherever else district politicians agreed to accommodate the programme.[18] The department also began to identify former plantation land on Java as an alternative for transmigrants who refused to depart again for what they called "the place of others".[19]

A second popular mobilization that was clearly linked to an ethno-territorial claim to the "right to exclude" was the eviction of more than 300,000 Madurese carried out by Dayaks in Kalimantan between 1997 and 2001. The Madurese were not official transmigrants. They had migrated to Kalimantan over a period of decades and had bought land to farm, laboured in the timber sector, or engaged in small-scale trade. In West Kalimantan, the Madurese comprised only 2 per cent of the population—hardly a numerical threat. Madurese were far fewer in number than the Javanese transmigrants who were untouched by the violence, and still remain in place even though their settlements were built on Dayak land.[20] In Central Kalimantan the Madurese made up at most 7 per cent of the population, concentrated mainly in two towns.[21] Thus, a struggle over land as a basis of livelihood does not explain why the Dayaks targeted the Madurese. Nor is jealousy over wealth a good explanation, since the Chinese, also numerous in these provinces and more prosperous than the Madurese, were left alone in this case, although they had been targeted in the 1960s during an earlier round of state-sponsored evictions.[22]

Dayaks who participated in or observed the evictions justified them in terms of the culture of the "offending" group: unlike the Javanese, Madurese arrogantly refused to accept their position as guests in Dayak territory, or to respect Dayak authority and jurisdiction. In interviews conducted by Nancy Peluso and Emily Harwell, Dayaks consistently referred to "territoriality or the territoriality associated with customary practice and Dayak identity".[23] Dayak determination to evict the Madurese on a permanent basis was reflected in a second round of attacks in West Kalimantan in 1999, attacks on camps sheltering displaced Madurese, and the murder of several Madurese who had returned to sell their property. In 2002, five years after the violence began in West Kalimantan, 78,000 Madurese were still in camps hoping for resettlement somewhere in the province. Dayak eviction of Madurese from Central Kalimantan in 2001 was more complete: 180,000 Madurese were quickly evacuated by ship to Madura and East Java, where 100,000 still awaited resettlement in 2005.[24]

The long stay of Madurese in the limbo of camps for "internally displaced persons" highlights an important fact: Following decades of voluntary and coerced migration, millions of Indonesians find themselves detached from the ethno-territories of their ancestors. In this case, Madurese sent back to their purported "homeland" encountered a cool reception. Unless they could find jobs or acquire access to a plot of land to plant, they were as vulnerable and as out of place in these homelands as they were in Kalimantan. To organize national space on the basis of exclusive ethno-territories raises the conundrum of exclusion in extreme form: some have their rights enhanced, while others have no place to go.

The second case we consider here, the 2001 mobilization of the indigenous Montagnards of Vietnam's Central Highlands to regain control over their ethnic homeland, has some similarities with the Indonesian case just described, but also some striking differences. In both countries, tensions were related to the arrival of large numbers of migrants under a combination of state-sponsored and spontaneous settlement. The goal of both mobilizations was to regain ethno-territorial sovereignty. The most striking difference was in the mode of mobilization itself. In Aceh and Kalimantan, large groups mobilized to undertake the violent eviction of the unwanted population. Rather than appeal to the government to solve the problem, they made good on their claim to ethno-territorial sovereignty by exercising it directly, in a visceral form that sent a clear message: We rule here; don't mess with us.[25] In Vietnam, in contrast, highlanders conducted peaceful mass rallies and presented polite petitions requesting that the government listen to grievances, honour previous promises and find appropriate solutions. Also different was the official response. In Indonesia, the perpetrators of the violent eviction and murder of the Madurese were not punished, and the "success" of the eviction was openly celebrated and upheld by senior Dayak officials who argued for imposing strict conditions on any Madurese who dared to return.[26] In Vietnam, the government suppressed Montagnard rallies with violence, and the organizers were killed or jailed.

Before we consider why the Montagnards' attempt to counter their exclusion took the form it did, we briefly recap how they became a minority in "their own" land, a story we told in Chapter 4. In the period 1975–2000, when the population of the highlands increased from 1.2 million to 4 million, the proportion of Montagnards dropped to one in four.[27] There were two migrant streams. One comprised ethnic Vietnamese brought in by the government to settle on state farms in designated New Economic Zones. The second comprised spontaneous migrants from the lowlands and from impoverished areas in the northern highlands who moved to the Central Highlands after economic liberalization in 1986. They came in search of land and were attracted by the coffee boom. Montagnards who no longer had sufficient land for swidden cultivation were compelled to intensify and joined the rush to plant coffee, but they were outcompeted by lowland Vietnamese who had more education, skills and access to credit. Many sold land to incoming migrants, while others had their land grabbed from them. Thus, the Montagnards were already facing a severe shortage of land when the coffee price crashed in 1999. Traders called in their debts, and more highlanders lost their land.[28]

The market was prominent among the array of powers deployed to exclude the Montagnards from access to land as a productive resource, but, as we noted in Chapter 6, market-mediated exclusion among smallholders is piecemeal and insidious, and it does not offer obvious targets for counter-attack. It is difficult to rally against market prices, or against migrants who have bought up land, one plot at a time. It was especially difficult for Montagnards to protest against practices in which they had participated, such as coffee planting, land selling and taking on debt. Mobilizing against the tangle of unjust programmes, laws and corrupt practices through which state agencies had appropriated large blocks of Montagnard land was also unfeasible, in view of the repressive stance of the ruling regime.[29] Instead, the Montagnards mobilized around a more general but deeply felt sense of loss—the loss of sovereignty over their homeland—and expressed their protest by respectfully drawing attention to promises that had been made to them by current or past regimes. Movement leaders referred to a grant of autonomy the French made in 1946, an edict signed by Emperor Bao Dai in 1951, and further promises concerning autonomy made by the Viet Minh in 1955 that were not honoured upon liberation in 1975.[30] An evangelical Christian movement led by a Montagnard who had moved to the United States helped to package the loss of sovereignty, the crisis over land, and the desperate struggle for livelihood into a unified protest movement. The Christians added an ethos of victimhood to the mix, as followers of the evangelical church were jailed and harassed for practising their religion.[31]

In 2001, thousands of Montagnards who gathered in a series of peaceful marches to present petitions demanding an autonomous homeland were dispersed by soldiers and police. Hundreds were arrested, and some were tortured. The government refused to acknowledge their grievances and accused them of being manipulated by foreigners seeking to destabilize the regime. Refugees fled to Cambodia, but the Vietnamese government insisted they were merely illegal migrants and pressed for their return. The entire highland population was placed under surveillance, their activities severely restricted.[32]

Although it ceded nothing on Montagnard aspirations for ethno-territorial autonomy, the government did recognize that Montagnards were facing a crisis of access to farmland. It proposed to release about 240,000 hectares of political forest for resettlement schemes, mostly in remote areas far from the centres of highland government where the Montagnards had hoped to strengthen their presence.[33] At the same time, it persisted with plans to bring in still more outsiders in planned resettlement schemes: 260,000 people by 2010, including 100,000 members of northern minorities who were scheduled for eviction to make way for a hydroelectric project located on the Black River (Song Da) immediately upstream of the Hoa Binh reservoir discussed in Chapter 5.[34] The government made no moves to restrict spontaneous migration, although it did attempt to address the problem of piecemeal land transfer by forbidding minorities from selling their land. The effect of this restriction, however, was mainly to drive land sales between highlanders and migrants underground.[35] Under conditions of land constraint, the Montagnards could not retreat into subsistence. They were compelled to participate in markets on adverse terms, in which market powers combined with their status as a "backward" minority made their hold on land chronically insecure.

The mobilizations to reclaim ethnic territories that we have examined in this section were attempts to counter a particular state strategy adopted by regimes in both Indonesia and Vietnam: the sending of members of national majorities into frontier zones as the "territorial spearhead" of states seeking to intensify control over people and resources.[36] This strategy deliberately undermined the autonomy of the ethno-territorial groups in situ, a move legitimated by state discourses that stress common citizenship (any citizen can live anywhere in national space) and the need to make full use of national land resources for the benefit of the people, especially landless people from the crowded agricultural cores. The only way to counter the first discourse is to reassert ethno-territorial sovereignty, a move accomplished violently in parts of Indonesia and not accomplished in Vietnam, where the Montagnard request for recognition was violently repressed and the migrant influx continued unabated.

Seen from a different perspective, however—the perspective of landless people desperate to improve their lot—government programmes such as the Indonesian transmigration programme, which enables landless people to move into less crowded territories and grants them farmland and initial support while they establish themselves, have met an urgent need. Some of the settlements fail on technical grounds, but some are successful, and they enable landless Javanese and Balinese families to gain land, livelihoods and sometimes prosperity as well. The transmigration programme continues to send out new settlers, often in conjunction with corporations developing oil palm. In Central Sulawesi, for example, 1,500 transmigrant households were scheduled to arrive in the period 2008–11.[37] But tensions with unwilling "host" communities over the farmland itself, or over the threat transmigration poses to the integrity of ethno-territories, make the land settlement solution to landlessness problematic. An alternative is for the government to redistribute existing agricultural land through programmes of land reform, but in an era of identity-based claims these programmes can also run into trouble, as we explain in the next section.

Land for Which People?

We introduced the topic of land reform in Chapter 2 and return to it here to consider how the ethno-territorial and identity-based claims we have just described have complicated, and to some extent undermined, leftist-inspired land-to-the-tiller campaigns for land reform. Our focus in this section is on Indonesia, where the rising prominence of identity-based claims has made the question "Land for which people?" much more fraught than it was in revolutionary Vietnam in the 1950s, for example, when the lines of transfer were clear enough: land was to be removed from landlords, both foreign and Vietnamese, and distributed to peasants.[38] Indonesia had a Communist-inspired land reform movement framed in precisely these terms in the early 1960s, but it was ferociously suppressed in 1965. By the time mobilization around land re-emerged in the late 1980s, the legitimating discourse had shifted to one that stressed cultural identity, environmental stewardship and attachment to place. Here we trace the reasons for this shift, and the consequences that follow from contending frameworks for legitimating claims to land.

The shift to a discourse of environment and culture was, in part, a strategic response to the repression of the New Order, which made any talk about land-for-the-people dangerous.[39] But there were other elements in play. One was a shift in the locus of popular mobilization from the agricultural cores of Java and Bali, where the demand had been for land redistribution, to the so-called outer islands, where the urgent issue in the 1980s and 1990s was the defence of people's existing land base from depredations by Suharto cronies who were extracting timber and other resources at a rapid rate. Corresponding with the shift in attention to the outer islands was a new configuration of allies. In place of the Communist Party, villagers confronted by attempts to exclude them from land and forests found support among legal activists and NGOs, many of them funded by donors with environmental mandates. These actors were eager to support the kinds of culturally distinct groups that were beginning to mobilize under the banner of indigeneity and assert rights to forests as a source of livelihood.[40] A movement framed in these terms mobilized with some success, to blockade logging and plantation roads and disrupt the bulldozers that were destroying farms. To a significant extent, Indonesia's indigenous peoples' movement found its constituency from among villagers who had mobilized to oppose state or corporate projects that threatened to exclude them from their ancestral land.[41] The movement was also linked, sometimes awkwardly, to bold claims to ethno-territories on a grand scale, notably Dayak attempts to reassert control over West Kalimantan and Central Kalimantan.[42]

In the last few years of the New Order, and especially after the fall of Suharto in 1998, the movement framed in terms of indigeneity and forest stewardship gained strength, as did the concept of ethno-territories defined in terms of belonging. During the same period, the peasant unions on Java and Sumatra reassembled and demanded implementation of the land reform provisions of the Communist-inspired 1960 Basic Agrarian Law, which had sat in limbo for three decades but had not been repealed. The main demand was for the distribution of public lands leased to plantations, more than 5 million hectares, especially areas that were unused or held under lapsed leases. The peasant unions also pressed for distribution of the political forest, a move that was relevant not only to landless villagers in Java and Sumatra, but also to farmers in the outer islands, where the state's claim to the political forest was the principal grievance. In Java and Sumatra, members of the peasant unions undertook what they called land reclaiming, or "land reform by leverage". This meant that they undertook mass occupations of thousands of hectares of plantation and forestland, which they distributed to members of the occupying group to farm, while seeking state regularization of their de facto possession and also pressing for legal reforms. The openness of these mass mobilizations was remarkable. Peasant union members proudly displayed union flags and carried union cards, apparently unperturbed by recollections of the 1965 massacre of organized peasants and the relentless intimidation of the New Order.[43]

In 2006 officials announced that the land reform promised in the 1960 Basic Agrarian Law would be revived and implemented through the distribution of 8.15 million hectares of "state land" currently under the jurisdiction of the Forest Department and the National Land Board. In the terms we introduced in Chapter 2, the announcement represented a slide from land reform envisaged as redistribution to the allocation of state-claimed land, and a further slide, in many cases, to the formalization of land occupations that had already taken place. It was also, in part, an attempt by state actors to reassert the state's right to regulate land access and effect some control over land occupations that had proven difficult to stop.[44] By 2010, delivery on the ambitious target had been negligible, however, and very little of the land claimed through the popular land occupations had been regularized by the granting of official licences or titles.

Having laid out the history of these distinct streams of mobilization around land, we can now look at what happens when contending frameworks for legitimating claims run up against each other. At the broadest level, arguments for indigenous people as environmental stewards, the integrity of ethno-territories, and land reform are congruent in that they all seek to defend people's right to land and promote social justice. But they differ in their answers to the question "Land for which people?" The indigenous rights movement supports moves to include indigenous people in management regimes for protected areas, but it rejects the idea that anyone else—landless people from elsewhere, for example—should be permitted to use parts of the land for agriculture. This difference was critical to the contest over land in the corner of Lore Lindu National Park. We introduced this case in Chapter 3, highlighting the tensions between villagers and the park. Here we take up another dimension of the story: the conflict that erupted among villagers pressing different kinds of claims to the land. When 1,000 families occupied the Dongi Dongi Valley inside the park, they justified their action in terms of landlessness. But their occupation was opposed by another group that made a counterclaim to the same land on the grounds of indigeneity and stewardship. According to the local activists supporting the indigenous cause, the proper solution to landlessness was not for landless people to take over the land of indigenous people, whose land rights were already precarious. Instead, they recommended, land occupations should be directed towards other targets—corporations, and corrupt officials who had appropriated land. They pointed to a nearby plantation as a suitable target—7,740 hectares, much of it unused, allocated to a Suharto crony. The people whose land had been taken by the plantation, however, disagreed: If the land was to be reclaimed from the plantation, they would do the reclaiming, and the land would return to them. A land reform activist countered that claiming land based on who lived where hundreds of years ago was symptomatic of Indonesia's growing "disease of origins" (*penyakit keaslian*) and was out of step with the common need of poor people (*rakyat kecil*) for land in a context where powerful people monopolized it.[45]

These events point to the conundrum—the unsolvable puzzle—that arises when indigeneity, ethno-territory and land reform, the three streams of counter-mobilization we have described in this chapter, run into each other and their grounds for legitimating exclusion collide. Since villagers everywhere in Indonesia contest the state's claim to land in the name of their customary (indigenous) rights, or their ethno-territorial rights, attempts by landless people to reclaim land from the state or state-sponsored corporations routinely come up against counter-claimants. Some argue that landless people have the right to "reclaim" land from plantations, or from the political forest, only if it originally belonged to them—if it was located in their ethno-territory or, more specifically, in their ancestral place. From the perspectives of indigeneity and ethno-territory, only the people with ancestral claims to land have the right to determine who should have access and who should be excluded. From the perspective of land reform, in contrast, the key to a just distribution is the need for land as the basis of an agrarian livelihood. Further, although land occupations are the leading edge of the land reform movement, the goal of the movement is to have the state follow through by recognizing the occupations that have already taken place, and making future occupations unnecessary because they are superseded by a proper state-run land-reform programme. In sum, they look to the government to set rules of access (and exclusion), acting on behalf of "the people" envisaged as a culturally neutral whole.

A surprise hitch that has arisen in the implementation of the Philippines' Comprehensive Agrarian Reform Program indicates that "culturally neutral" citizens who are free of extraneous attachments may be illusive in that context too. A study of conflicts "among the people" on a plantation slated for reform found that retired workers, and children of workers who had grown up on the plantation but were currently living and working elsewhere, argued for a share of the land based on their attachment to the plantation and to the communities they had built there. They rejected the idea that they should be excluded while current estate workers, some of whom were newcomers, were given land. In this case it was not cultural distinction, but duration of attachment and the way the plantation had become imbued with their sweat and toil, and their personal histories and relationships, that gave them a sense of entitlement.[46]

There are two main criteria by which land reform activists in Indonesia identify the people they consider legitimate beneficiaries: they should be landless, and they should be committed to farming the land they acquire. The second criterion relates to one of the legitimating discourses for land reform, that is, the link activists draw between the right to land and the right to food, which falls under a UN guarantee. The imagery associated with this right is the self-provisioning peasant household and community, an image promoted by the Indonesian Federation of Farmers' Unions (FSPI), which is closely linked to the transnational NGO Via Campesina and the Food First Action Network. FSPI rejects export agriculture on the grounds that smallholders run the risk of losing their land and livelihoods when the costs of production exceed the returns[47]—a problem we outlined in Chapters 4 and 6. Nevertheless, some smallholders prosper when they ride the wave of a commodity boom, and many would like to try their luck. So the attempts by movement leaders to protect smallholders from the risk of intensified market engagement and promote an "alternative to capitalist rural society" based in "eco-collective farming" on jointly held land are not necessarily welcomed by the groups actively occupying land.[48] Even if they share the desire, it is highly unlikely they can put it into practice in a context such as rural Java, where, as we have seen, market relations have been entrenched for at least a century and withdrawal into subsistence production is not an option. One study of a land-reclaiming site on Java five years on found that two-thirds of the farmers who initially occupied the land had sold up and left the site.[49] More studies will be needed to establish why that was so: whether the people never intended to farm the land but wanted to sell it to finance other ventures, or whether they tried to farm it but were unable to retain it, for the "everyday" reasons we discussed in Chapter 6. In a fast-changing rural scene, securing access to land at one point in time does not mean holding the land indefinitely. The same is true when land is reclaimed on the basis of belonging: the people asserting their right to ethnic homelands often draw their livelihoods from multiple sources, in which a patch of land on which to farm, or to build a house for retirement, comprises only one strand. If the question "Land for which people?" were answered in favour of subsistence farmers, the number of qualified applicants would be few indeed.

Against Eviction

This final section examines large-scale mobilizations by people resisting eviction related to state or state-backed schemes such as conservation-inspired enclosure of forests, and the building of infrastructure, especially dams. People facing eviction are usually clear about what they reject—the loss of their communities and livelihoods. Scaling up the struggle and finding allies, however, requires diverse affected communities to cohere around a positive programme. We draw on the case of Thailand's Assembly of the Poor to ex-plore the discursive and organizational strategies used to transform site-specific resistance into a broad social movement.

On-site eviction has a special capacity to unite resistance across class and ethnic lines, because the form of exclusion is both direct and catastrophic, and large numbers of people are threatened simultaneously by a single source. On-site resistance is vastly strengthened, however, when its legitimizing discourse appeals to national and transnational constituencies concerned about issues such as human rights, indigenous people and the environment. Increasingly, such concerns are articulated by networks of student activists, NGOs and donors that help to draw connections between local events and global problems. Often, mobilization is bolstered by outrage at the violence associated with large-scale development schemes, and compounded by corruption in their implementation. The moral clarity of discourses that highlight unjust treatment of people who are clearly identifiable as victims means that activists and journalists are quick to rally to the cause. But when the rights and wrongs of exclusion and access are viewed across scales, moral clarity can dissipate, especially in the case of hydro dams: people on site may lose access to land and livelihoods, but farmers downstream gain access to irrigation, and the children of farmers gain access to factory jobs fuelled by cheap electricity. In this matter, as with the others discussed in this chapter, the conundrum of who must be excluded for others to have access is fraught with the potential for conflict among social groups with different interests and claims.

In Indonesia, it was opposition to dam construction at Kedung Ombo in the mid-1980s that renewed the links between urban activists and farmers that had been cut off with the massacres of 1965.[50] In the Philippines, opposition to the Chico River Dam in the 1970s played a similar role. The scaling up of dam-based protests was most successful, however, in Thailand, where in 1995 a series of local mobilizations provoked mainly by eviction were stitched together into a national movement of people who found common ground as "victims of development". The movement called itself the Assembly of the Poor, a label that reflected its organizational structure (an assembly of equals, a conference) and served as a reminder of the gap between rich and poor in

Thai society.[51] The fact that *sammacha*, the word translated as "assembly", is used in Thai to refer to party congresses in socialist countries gives a sense of some of the resonances consciously evoked by the movement. The Assembly's legitimating discourse blended this set of associations with others drawn from the repertoire of ethnic minorities and indigenous people, the iconic victims of development who seemed to offer the promise of more equal and sustainable development alternatives.

In 1997 the Assembly mobilized 25,000 people to form an encampment in the heart of Bangkok that extended for 99 days and attracted sympathetic media attention.[52] The Assembly presented a long list of grievances from its constituent local groups and sought site-specific solutions as well as broad policy reforms. All the grievances had eviction at their core. The Forest and Land Group was made up of 93 grievances affecting about 22,000 families. It was centred in the northeast, where millions of farmers who had been encouraged to move into upland forest reserve areas in the period 1960–80 to help counter the Communist insurgency were suddenly declared squatters on state forestland.[53] In 1991 the army launched a programme to evict them, which would, if fully implemented nationwide, have relocated 5.8 million people (see Chapters 2 and 3). The Forest and Land Group also included minorities such as Karen from the north, similarly threatened with eviction, in this case mainly from recently declared national parks. The demand from this group was for legal rights to continue cultivating on state-claimed forestland. Second in size was the Dams Group, with 16 cases involving 11,000 families seeking the halting of planned dams or compensation for eviction caused by dams already completed. The focus was again the northeast, where most hydro projects are located. Two smaller groups were State Development Projects and Slum Community, together representing 3,000 families in 13 sites also affected by evictions justified by the state in terms of development and the common good.[54]

Two generations of activists supported the local groups and helped formulate the broad arguments in the Assembly's political programme. Some of these activists had been members of the Communist Party of Thailand (CPT) or the student movement that was brutally crushed in 1976, and had spent up to a decade in forests and remote villages with the CPT. When a 1980 amnesty enabled them to return to urban life, some became active in development-oriented NGOs, a newly emerging organizational form that enabled them to pursue their commitment to rural people suffering the effects of inequality while adding a commitment to social democracy and non-violence, platforms absent in the CPT. The second set were younger activists recruited to the NGO movement at a time when the ecological damage of the push for "development" was becoming apparent, together with the high price paid by the rural poor.[55]

As was also the case in Indonesia, environmentalism had served from the early 1980s as a convenient proxy for opposition at a time when land-based mobilization was still labelled Communist and hence subversive.[56] There was also a third set of activists, people from the villages and small towns of the northeast who had acquired national perspectives through education and labour migration, who returned to their home regions to help defend villagers against threats to their land and livelihoods.[57] Together, these activists built networks among villagers and linked their predicament to forms of state-supported development that secured the wealth of the urban elite at their expense.

The Assembly framed its alternative model of development in terms of "community-based resource management" and the culturally distinct "local knowledge" associated with the "traditional" people and indigenous hill tribes, assumed to have remained close to nature. The Assembly was clear about what it opposed: large-scale development projects, and the greed of the national elite, based in Bangkok, and its transnational allies. It was implicitly anti-capitalist in its emphasis on subsistence production and sustainable resource use, although the point was not elaborated. Its key demand was for the participation of people in decisions that affected their livelihoods, and respect for human rights.[58] In crafting its image, the Assembly glossed over the capitalist character and class divisions of rural society. It understated the mobility of its rural constituents—especially those of the northeast—who combined labour migration to factories in Bangkok and provincial towns with agriculture in multi-stranded livelihood strategies.[59] It used symbols that promoted solidarity within the movement and appealed to the sympathies of the urban, educated, media-watching public.

The striking feature of the Assembly's protracted mass action in 1997 was how little it was able to accomplish. Backed by the presence of a massive and tenacious crowd, the Assembly's representatives negotiated for weeks with politicians and bureaucrats to secure resolutions to their grievances. They succeeded in extracting significant promises on the issues of compensation for people displaced by dams and other infrastructure, and the right for long-settled groups to remain in forests.[60] A few months later, however, the Asian economic crisis hit and a new government reneged on all the major agreements, clamped down on protests, reasserted the paternalistic wisdom of bureaucrats, and pressed ahead with its programmes. Local groups continued to take local action—blockading dam construction, resisting eviction—but the momentum achieved by the Assembly was difficult to sustain. State-supported violence against on-site resistance increased. Officials revived the accusation that the protesters were Communist, a move intended to intimidate them. They also stepped up the long-standing accusation that the Assembly was not driven by the authentic concerns of villagers but rather by NGOs that were overly influenced by transnational donors. NGO offices were raided.[61] Yet the authorities did pay attention to the scale of the protests as a sign that future attempts to impose mass evictions would be contested, and quietly shelved some of the projects they had planned.

Despite its populist stance and initial overtures to the Assembly of the Poor, the Thaksin government (2001–6) also failed to follow through on the Assembly's major demands. Yet villagers in the north and northeast, the strong-hold of the Assembly, voted three times for Thaksin. They were especially posi-tive about his programmes offering rural credit and village-level block grants that moved away from paternalistic guidance to treat them as competent actors, capable of managing local affairs. These voters sought a closer relationship with the state and a boost to their micro-capitalist enterprise—a far cry from the community-based and subsistence-oriented "alternatives" the Assembly proffered as a counter to large-scale development schemes. Their "betrayal" of the movement, Andrew Walker argues, sent an important message: activists concerned about rural livelihoods needed to pay more attention to the complex engagements between rural people, the capitalist economy and the state.[62] The rural people who mobilized in the Assembly rejected the militarized, state-based powers of exclusion arrayed against them, and they most definitely rejected schemes that evicted and impoverished them, but through their support for Thaksin they indicated that they had no objection to capitalism or modernity as such, and certainly no quarrel with the idea of "development". Their demand was for inclusion in the national project, based on transparent planning processes and fair access to resources. The protracted occupation of prime commercial locations in Bangkok in 2010 by thousands of Red Shirts supporting Thaksin sent this message even more sharply. Their demand was for a place in the nation as citizens with fully fledged democratic rights, to be exercised across the spectrum of decision-making arenas in the rural hinterlands, and in Bangkok, where they sought to compensate for their exclusion from the ballot by a visceral presence in the seat of power.

Conclusion

Like our other chapters, but more sharply, this chapter has highlighted exclusion's double edge. We have explored the conditions under which people have mobilized collectively to counter their exclusion from land as territory or productive resource, and sought to exercise, in turn, their own powers to ex-clude. The underlying conundrum—that people want the right to exclude, but don't want to be excluded—came up in each of the mobilizations we examined:

the struggle of indigenous people to hold on to their land, struggles to restore ethno-territorial sovereignty, programmes for land reform, and resistance to eviction. Struggles erupted not just at the level of distribution—who should have what share—but over which of several contending frameworks (stewardship, belonging, need) would prevail.

The most obvious power at work in the counter-exclusions we examined was legitimation, as different parties embroiled in struggles over land and territory sought to mobilize allies and explain the justice of their cause. Force was also prominent, as a power used to evict people who were—from one perspective or another—out of place. Countervailing force also lay behind the power in numbers that confederations such as the Assembly of the Poor brought to their protests in the national capital. Mobilization was most difficult when the market—routine capitalism—was a key exclusionary force, the situation for highlanders in Vietnam caught up in coffee's boom and crash. While Communist movements had a principled stand against the operation of market forces, for which they substituted state planning and collective production, current mobilizations for access to land often have a markedly anaemic analysis of capitalism, as if everyone granted access to land could and would stay permanently in place. Explicit attempts to protect specified groups of citizens from the operation of a land market by making tenure collective and/or inalienable—a pattern that is emerging throughout Southeast Asia, as we showed in Chapter 2—run into the problem that making land sales illegal does not stop them from happening. Yet despite the challenges, all over Southeast Asia people struggle for land and hold on to it as long as they can. Often, they put their lives at risk. They do so because, even in a context where livelihood strategies are diverse, they are very much better off with access to land than they are without. Land remains a critical livelihood resource. Land as territory, a village, a place to call home, turns out to be equally important: people unable to assert an effective claim to place are vulnerable in the extreme and sometimes end up, like Madurese evicted from Kalimantan, in the nowhere land of a displaced persons' camp.

Notes

1. Lucas and Warren 2003: 99.
2. O'Brien 1996 proposes the label "rightful resistance" for struggles framed in these terms. Labbé 2010 uses this concept to examine resistance in peri-urban Hanoi. See also Nguyen Van Suu 2007.
3. See the discussion in Li 2000, 2010b.
4. Li 2001a, 2007b, 2007c discuss this point in the context of Indonesia.

5. Brown 1994, Wenk forthcoming, McDermott 2001, Perez and Minter 2004, Dressler and Turner 2008, Hirtz 2003, Yap 1998.

6. Winichakul 1994. See Delcore 2007 and Wittayapak 2008 for analyses of the purported differences between rural Thai, who are viewed as uncivilized but improvable, and intractable tribes.

7. Colchester n.d.; Forsyth and Walker 2008: 60–1.

8. Vandergeest 2003b: 24, 26.

9. Walker 2001; Delcore 2007: 101–2. Note that the Karen were considered Thai until the 1960s, as the term "hill tribe" was reserved for groups at higher elevations. They were relabelled a hill tribe on the grounds of their swidden practices inside newly mapped forest boundaries. See Vandergeest 2003b: 24, 26.

10. Delcore 2007: 100.

11. Vandergeest 2003b: 27; Wittayapak 2008.

12. Vandergeest 2003b, Forsyth and Walker 2008.

13. McCarthy 2004; International Crisis Group 2003; Benda-Beckmann and Benda-Beckmann 2001; Klinken 2007, 2008.

14. See also Warren's discussion of the response to tourist development of villages in Bali, discussed in Chapter 5.

15. *Kompas*, 29 July 1999, 18 Nov. 1999, 20 Nov. 1999.

16. A year later, 23,000 people were still being supported by the Social Affairs Department in North Sumatra, the rest having returned to Java (*Kompas*, 24 Oct. 2000).

17. *Kompas*, 20 Nov. 1999, 16 Oct. 2000a, 16 Oct. 2000b.

18. Down to Earth 2001b: 6–7.

19. *Kompas*, 16 Oct. 2000a, 16 Oct. 2000b.

20. Peluso and Harwell 2001: 95.

21. International Crisis Group 2001: 1, 14. See also Down to Earth 2001a.

22. Peluso 2008 discusses the link between these two sets of evictions.

23. Peluso and Harwell 2001: 108.

24. For updated data as of 2009, see http://www.internal-displacement.org/8025708 F004CE90B/httpDocuments/82772DA861FE558CC12575840055CCF2/$file/ West+and+Central+Kalimantan.pdf.

25. Klinken 2008 argues that Dayak politicians deliberately used theatrical tactics against a relatively weak group precisely in order to send this broader message.

26. International Crisis Group 2001: 22.

27. De Koninck and Dery 1997; De Koninck 2006.

28. This history is summarized in Human Rights Watch 2002.

29. Human Rights Watch 2002: 10.

30. Human Rights Watch 2002: 17–25.

31. Human Rights Watch 2002: 9.

32. Human Rights Watch 2002: 83, 90, 91, 94.

33. Phan Trieu Giang 2010.

34. Human Rights Watch 2002: 40, 82, 121.

35. Phan Trieu Giang 2010.

36. De Koninck and Dery 1997.

37. http://www.depnakertrans.go.id/microsite/KTM/?show=p15.

38. Kerkvliet 2005. The occlusions that were necessary to produce the apparently homogenous category of "peasant" are discussed in Tsing 2003.
39. Afiff and Lowe 2007.
40. Lucas and Warren 2003.
41. Li 2000, 2007b, 2007c.
42. Li 2002b, Davidson 2008.
43. Lucas and Warren 2003, Afiff *et al.* 2004, Lapera 2001.
44. Bachriadi and Sardjono 2006.
45. For a full discussion of this case, see Li 2007c, Chapter 5. See also Sangaji 2007. Tensions between different groups asserting the right to reclaim plantation land in Sumatra are noted in Lucas and Warren 2003: 93n33.
46. Rutten 2010.
47. La Via Campesina and Food First Information and Action Network n.d., Federation of Farmers' Unions n.d.
48. Bachriadi and Sardjono 2006: 27–9.
49. Lucas and Warren 2003: 92n30, 94n36. See also Bachriadi and Sardjono 2006: 27.
50. Aditjondro 1998.
51. Missingham 2003: 44, 60; Baker 2000: 16.
52. Missingham 2003: 2–3.
53. Missingham 2003: 34–6; Baker 2000: 14.
54. Missingham 2003: 44–50.
55. Missingham 2003: 97–120; Baker 2000: 18.
56. Hirsch 1993.
57. Baker 2000: 10–1.
58. Missingham 2003: 54–9.
59. Baker 2000: 7–10; Walker 2007.
60. Baker 2000: 23.
61. Missingham 2003: 99, 201–9; Baker 2000: 24–8.
62. Walker 2007. Walker 2009 describes Thai farmers' pragmatic experimentation with contract farming as a way to plant new crops without risking their own capital.

8

Conclusion

'Cause when love is gone
There's always justice
And when justice is gone
There's always force
And when force is gone
There's always Mom
Hi Mom!
       – Laurie Anderson, "O Superman"

This book has traced key processes transforming how land is used in Southeast Asia, and the ways people can access or be excluded from land. Since 1990, licensing regimes have intensified; environmentalism has become ambient; volatile crops have expanded the agricultural frontier; post-agrarian land uses and livelihoods have changed the face of the countryside; intimate relations among villagers have fractured and realigned; and groups have mobilized to counter the powers deployed to exclude them, and to implement exclusions of their own. Our goal has been to explore how these processes unfold, and the dilemmas they present for differently situated actors.

Exclusion from land, we have argued, is not new. Nor is it a problem to be corrected, since every productive use of land requires exclusion. But the rapid and multifaceted transformations of the nature of exclusion in Southeast Asia, and the extent to which these changes will have permanent consequences, make the *how* and the *who* of exclusion absolutely central to understanding the reconfiguration of wealth and power in society. We have approached this question by focusing on the powers deployed. Like Laurie Anderson in her 1981 minimalist new wave hit "O Superman", we have identified four powers shaping social relations. Our list—not identical to hers—emphasizes powers of regulation, the market, force and legitimation. And as we have shown, these powers do not so much follow as reinforce each other.

Our first step towards drawing conclusions from our study is to sum up what we have learned through our focus on powers. We then draw out the dilemmas that lie at the heart of this book and summarize how these dilemmas are addressed by a broad assemblage of actors who seek to affirm land's social function but run up against the stubborn fact of exclusion's double edge. Finally, we use the food and economic crisis of 2007–9 as a point of departure for considering the implications of our analysis for agrarian and post-agrarian futures in the Southeast Asian region and beyond.

Summing up the Powers

The first of our four powers, regulation, involves setting rules that delimit land's boundaries and shape how land can be accessed and used, and by whom. The most important regulatory actor is the state, and we argued in Chapters 2 and 3 that state efforts to regulate land are central to processes of land titling, land allocation, land formalization, and conservation. In each of these cases, state actors are concerned to define plot boundaries and territorial boundaries by surveying and zoning. States designate the use of some land for agriculture, some for conservation, some for production forest, some for tourism and so on. States also seek to regulate who can access land, for what purposes, and under what conditions. Through titling programmes, states intervene to shift the grounds of access from social identity to market power. States regulate who is set free to dispose of land as a commodity, and who is constrained. Regulated limits to land sales and land uses are most obvious with respect to land reform and resettlement programmes, but conservation concerns can also trigger stringent conditions on access and use. As we showed in Chapters 3 and 7, villagers who can make a claim to being proper environmental stewards have a better chance of receiving state permission to live in areas zoned as forest.

The state power to regulate also facilitates, and sometimes drives, other processes we have examined. Producers of boom crops often receive land through state allocation, and smallholders who fail to take up a boom crop—rubber in Laos or oil palm in Borneo, for example—can find themselves excluded by state regulations favouring "productive" land use. Throughout the region, states prohibit swidden on these grounds. States also try at times to hold back the spread of booms, as we saw in the case of efforts by the Thai government to keep shrimp farms out of mangrove forests and inland areas. State planners are centrally involved in land conversion, as they rezone agricultural or forest land for industrial, commercial, residential or tourist developments, and state regulatory power is critical in land-hungry infrastructure projects such as large dams. States allocate land to their component ethnic groups on the basis of different criteria. Moving further from direct regulation of land, state subsidies that reduce the prices of agricultural inputs for selected beneficiaries, and state investments in infrastructure such as irrigation and roads, play a key role in shaping "intimate exclusions", as we saw in Chapter 6.

None of this is to claim that state actions are always effective, or that states are coherent, unitary actors. Nor does it suggest that states are the only rule makers with respect to Southeast Asian land. Non-state actors such as NGOs and donor agencies increasingly shape the rules that govern land access and exclusion, most notably in the field of conservation. Farmers, too, can at times work together (though not in isolation) to set some of the rules governing land access and exclusion in their villages, a dynamic we explored in the case of community-based natural resource management in Chapter 3, and rules against land selling devised by villagers in Laos, Bali and upland Vietnam (Chapters 2, 5 and 6). The power to make the rules, then, is vital both to the process of exclusion and in determining who is excluded.

The second power we emphasize is that of the market. Markets are complex, socially embedded institutions that are underpinned by regulation, legitimation and force. Yet one form of the market in particular—its manifestation in the price of land, or crops, or agricultural inputs, or labour—confronts farmers as a disembedded but highly persuasive social fact. In Chapters 4 and 6, we saw high global prices for crops such as cocoa, coffee, oil palm and shrimp pull millions of Southeast Asians into boom crop production, with profound implications for exclusion as migrants flooded uplands and coasts to join the boom, as states and corporations grabbed land and states rezoned it, and as the people who had held land before the boom took steps to make more individualized, and less fuzzy, their claims to it. Prices also shaped exclusion on the way down, as we saw in the case of the coffee bust of the early 2000s that intensified land loss among minorities in the Central Highlands of Vietnam. The cases of land conversion to non-agricultural uses highlighted in Chapter 5, on the other hand, saw prices pushing people away from agriculture rather than pulling them towards it, as skyrocketing land prices encouraged farmers to sell either their land or their tenant rights, or to try to hold on to them for investment or speculative purposes.

Both Chapters 4 and 5 showed very clearly that rising prices do not only change market dynamics, they also prompt a wide range of actors to mobilize the other powers of exclusion at their disposal to lay claim to newly valuable land. We saw something similar at a much more intimate scale in the Sulawesi highlands, where individuals acted to enclose ancestral land to the exclusion of their kin (Chapter 6). Our discussion of "everyday" processes of accumulation and dispossession took up classic themes in agrarian political economy by emphasizing that farmers struggling for survival amidst a wide range of price signals (including prices for land, for crops, for labour, for borrowed money and for agricultural inputs) are subject to a "simple reproduction squeeze" that pushes many of them out of agriculture, while the social ties that might protect them from this process are more fragile than discourses of shared poverty and village solidarity suggest. In Chapters 2 and 3, finally, we saw a range of actors—most notably states, but also multilateral lenders and NGOs—trying to call *new* markets and price signals into being and to make them work for development purposes, the most prominent being the market for land in Chapter 2 and that for "environmental services" in Chapter 3. This desire to align price signals with development objectives, a signature move of neo-liberal policy, is having an increasing impact on exclusion in Southeast Asia. Yet it remains contingent on powers of legitimation proposing, for example, that transferable property rights are a sine qua non for development, and on powers of regulation in forms such as Land Law, which backs up the rights of owners and displaces more local recognition of who gets to exclude whom from land.

Force, the third of our powers, is in some ways the most straightforward. Contrary to the images conjured up by anti-land-grabbing advocacy, justified though it may be in many circumstances, brute force is not the dominant power in any of the processes we examine. Yet it is very much present in all of them. In our discussion of conservation, the main use of force was by state actors seeking to expel villagers from their homes inside conservation areas, but we also noted the distressing case of ethnic Thais from the lowlands joining forces with an NGO and relying on tacit support from some local officials to burn down the orchards of upland Hmong who were allegedly degrading the local watershed through deforestation and chemical use. Boom crop expansion, often a "frontier" phenomenon, has long been associated with violence, as state and corporate actors grab land, migrants seize fallows from swidden farmers, and people without the ability to resist overtly use "everyday" forms of violent resistance such as arson. We also noted the striking fact that peri-urbanization, a process that ties Southeast Asia more deeply into the global economy, often takes place in a pervasive atmosphere of fear, intimidation and coercion— indeed, that peri-urban areas are in many respects just as lawless a "frontier" as the uplands and coasts where boom crops are grown. Violence stood out as a component of the land occupations described in Chapter 7, especially in the cases of ethno-territorial mobilization that saw different groups using force to expel unwelcome "outsiders", with state actors at times sitting back (as with respect to Dayak attacks on Madurese in Kalimantan) and at times clamping down on these mobilizations with brutal force (as in the Central Highlands of Vietnam). Arson reappeared as a tool of intimate exclusion in the Sulawesi highlands, and violence was a feature of struggles among landless agricultural workers in Java trying to defend the territories where they had privileged access to jobs.

Legitimation, finally, is vital to exclusion. It has power in itself, in that at times people will relinquish or allow claims to land on the basis of the compelling power of discourses of right and wrong. But it also provides the indispensable normative underpinning to rules, rights to buy and sell, and violence that makes them seem legitimate or, in some cases, makes them so much a part of the natural order of things that they are not up for debate or analysis. One of the most important forms of legitimation we explored in this book is the idea of territory, or the association between land and belonging. Across Southeast Asia, people argue on the basis of nationality or ethno-territorial belonging that their group—or, in some cases, they personally—have a right to land. Nationality pulls in different directions here. States claim that they must guard, regulate and allocate the nation's land for the greater good; landless people see access to land as a citizenship right; and in places such as Aceh, Mindanao and West Papua, alternative conceptions of nationality are articulated against the dominant ones. Claims to govern territories based on ties of belonging run counter to the state right to define land uses or—more contentiously—to redistribute land on the basis of need.

The legitimation supplied by visions of development, modernity, civilization and what we have called ambient environmentalism has a pervasive effect on land use and exclusion. These dreams may promote the expansion of agriculture, as when boom crops such as oil palm and shrimp are put at the centre of development plans; or they may push in the opposite direction, as "backward" agricultural land uses are displaced by industry, commerce, housing, tourism and infrastructure. Even at the intimate scale, as we saw in Chapter 6, visions of development were deployed to legitimate exclusion among neighbours and kin, and to counter alternate discourses of mutual obligation and care. In the arena of conservation, too, different legitimating conceptions push against one another, with one set of disputes pitting conservation against other land uses, and another pitting different approaches to conservation against each other, often with the question of exclusion at the centre of the debate. Finally, the ideological power of markets as means of improving livelihoods, promoting economic growth and conserving nature is on the rise everywhere in Southeast Asia, legitimizing programmes that enhance the alienability of land and the buying and selling of land-based resources to those offering or taking the highest price. This ideological power of the market extends to buying and selling rights to influence the use of land and land-based resources to supply "environmental services".

We have made the case in this book that an analysis based in these four powers can shed a great deal of light on dynamics of exclusion, both in Southeast Asia and elsewhere. We do not wish, however, to push this emphasis on powers too far. Two qualifications of our argument, in particular, are crucial. First, as we have noted many times, the separation of these powers from each other is an analytical and heuristic move. We argue that distinguishing these powers is helpful for understanding how exclusion works, but we do not claim that this separation exists in practice. Rather, the four powers we have highlighted operate together, and in many cases, as we state most explicitly in Chapter 5, they are inextricably fused. Second, our choice of these four powers is not an assertion that they are the only ones that matter. We highlight regulation, the market, force and legitimation in part because we think them extremely important, but also because a larger array would have made the analytical framework unwieldy. It is clear throughout the book, however, that other powers are in operation. Four of these can be briefly mentioned here. First, we have often seen *environmental change* working to exclude people from land access, for instance, through the salinization and pollution associated with shrimp farming, the environmental consequences of peri-urbanization for farms that remain in areas undergoing conversion (including greater infestations of pests such as rats), and the landslides that hit cocoa farmers in Sulawesi as they moved onto progressively steeper land. Second is growth in, and control over, new *knowledge and technologies* for crop production, transportation, communications, remote sensing and so on that influence both the incentives for new forms of exclusion and capacities for monitoring and enforcement.[1] Third, *political relationships and alliances* are critical to land access and exclusion. State actors may award land to supportive groups, and NGOs and peasants' movements try to work together to increase their leverage in fights over land. And finally, as we noted in the introduction, *inertia*—the force of what exists— has a power of its own. The use of other powers of exclusion is not costless and carries risks, so existing relationships may continue simply because changing them is too difficult or too uncertain.

Land Dilemmas

Changing relations of access and exclusion present dilemmas not only for the people directly involved as actual or potential land users, but for those who take on responsibility for devising the arrangements that determine exclusionary outcomes. These actors include politicians and government officials; the national and transnational experts who advise land agencies; scholars, advocates and activists lobbying for changes in policies or practices; and social groups that mobilize around their own visions of how land should be used, by whom, and who should decide. We have argued throughout that analyses that focus on exclusion as a "good" or "bad" thing are likely to go astray as guides for making land access in the region more just, since exclusion itself is an unavoidable fact of land access and use. The question of exclusion needs to be drawn, rather, in terms of what different exclusionary forms mean for different groups and societal outcomes. Some analysts, such as the de Soto-inspired neo-liberals we discussed in Chapter 2, argue that one format—exclusive private property held under fully documented title—will be good for everyone, increasing investment in rural production, including rural people as full market subjects, and deepening democratic citizenship. In making this argument, they ignore or understate the fate of people who lose out as a result of titling schemes. Some critical scholars, on the other hand, see any form of private property—even forms that are clearly desired by millions of people across the region—as "written in letters of blood and fire" and argue against exclusion on these grounds. In contrast to these positions for or against exclusion, our analysis focused on exclusion's double edge. Because all productive land uses require exclusion, the critical issue is who will win, and who will lose, from the ways in which boundaries are drawn.

An important finding from our study is that the dilemma presented by exclusion's double edge is not hypothetical, and it is not restricted to a narrow group of policy wonks: it is widely recognized in land debates and practices throughout the region, it showed up repeatedly in the material we examined, and it presents itself to actors at many different levels. In Chapters 2 and 3, we saw the emergence of hybrid tenure regimes that attempt to balance individual land uses with restrictions governed by environmental concerns (rights to use sloping land in the political forest should not take the same form as rights to flat land in the agricultural core). Other restrictions seek to keep market forces at bay. Beneficiaries of land reform programmes in core agricultural areas may be denied the right to sell their land, both because the grant was made in order to solve the problem of landlessness, and because the authorities designing land reform schemes know all too well how quickly market powers can dispossess people. These restrictive schemes, however, are very difficult to enforce, and usually their main effect is to drive land sales underground, as we saw in the case of state efforts to protect indigenous land in the Central Highlands of Vietnam. The more effective examples were based on initiatives from below, supported by an enabling legal framework, as in some (though not all) of the villages we described in Laos and Vietnam (Chapters 2 and 6). But communities—even when closely integrated by relations of kinship—also use market mechanisms to exclude each other, a long-standing feature of village Java, as we showed, and a feature that emerged with striking rapidity in upland Sulawesi.

We saw dilemmas of exclusion emerge in other processes too. State actors supported boom crop production but worried about food security and conservation; conservationist NGOs wanted protected areas but knew that the burden of creating them fell on local villages; farmers wanted title to their land but saw that receiving it meant paying more taxes or more easily losing their land when times were tough. The most striking and intractable expression of exclusion's double edge emerged from our discussion of identity-based and ethno-territorial claims to land and territory, in which the use of belonging as the principal criterion of entitlement clashed awkwardly, and sometimes violently, with other rationales.

The comparative sweep of our book, and the focus on broad processes and policy trends along with specific national and local examples, also reveals how land dilemmas are understood and worked upon differently in various contexts. This is not to suggest that each case is too different from the next to draw out patterns. Rather, it shows the need to situate dilemmas—real and difficult policy, advocacy and action choices—within the specific historical, geographic, social, economic and political contexts that frame them, in order to analyze them in ways that yield some insight into the factors at work at different sites and conjunctures. In a situation where there is already a cacophony of actors advocating different positions and offering policy advice, we do not propose to offer our own version of the optimal arrangement. Contentious land issues, as we have shown, are resolved not only in the arena of policy and regulation, but viscerally through force, and through the logic of the market. Exploring how these powers work to shape outcomes on the ground is our principal contribution.

Stepping back, however, we have to recognize that some of the dilem-mas we have explored really are intractable. They are conundrums—puzzles that tease but cannot be resolved. In his brilliant analysis of Goethe's *Faust*, Marshall Berman argued that the recognition that it is not possible to create something without destroying something else is close to the core of modernity. He pointed out that Goethe has Mephistopheles tell Faust, in what may be history's most extreme expression of the "no omelettes without broken eggs" concept, that even God's creation of the world "usurped the ancient rank and realm of Mother Night" (that is, in order for there to be Something, Nothing had to be destroyed).[2] This does not mean, however, that everything is a dilemma for everybody all the time. When farmers are evicted from their land, by force and without compensation, to make way for a dam, there is no dilemma for them. To claim that exclusion is often riddled with dilemmas is not to argue that we should throw up our hands and abandon efforts to determine who wins and who loses. But we do suggest that neither focusing on the magic of land titling and land markets nor assuming the presence of collectively oriented peasants who are united in their resistance to capital and enclosure is as likely to produce sustainable or socially progressive outcomes as more qualified and nuanced analysis.

From the Crisis to the Future

The linked food, financial and economic crises of 2007–9 serve as a useful entry point for considering the implications of our analysis for the future of land in relation to livelihoods for the hundreds of millions of people who continue to live and work in rural Southeast Asia. Most of the empirical material presented in this book is taken from studies carried out between 1990 and 2006 (though we have situated them in relation to patterns of agrarian change from earlier eras). While we did not have the sense while designing the framework of the book that we were describing a particular period, these years—covering as they do the decades of "globalization" between the end of the Cold War and the global economic crisis that began in 2007—are increasingly coming to be seen as a specific era in the history of the global political economy. What comes next is not yet clear. What we do know is that in 2007–9, Southeast Asia, like the rest of the world, was buffeted by a series of interconnected and wrenching crises in markets for agricultural and other commodities, and in global finance, production and trade. When we turn the tools of analysis we have developed in this book towards analysis of that crisis, some disturbing features come to light.

The most obvious and troubling effect of the crisis on agrarian Southeast Asia was the terrible pressure it put on rural livelihoods. In 2007 and early 2008 (that is, after the global financial crisis had begun, but before its effects spread into global stock markets), there was a worldwide spike in the price of key agricultural commodities. Rice—the most important food crop in Southeast Asia—was affected more than most crops by this spike. The price of Thai rice rose 255 per cent between 2004 and 2008, with most of the increase coming after mid-2007. By early 2008 the Southeast Asian market was in a state of panic as key rice-exporting countries (including Vietnam) restricted exports, and importers such as Malaysia and the Philippines scrambled to cover their needs.[3] While prices fell back in mid-2008, they remained high in historical terms. In terms of the concerns discussed in this book, the most immediate impact of high prices for rice and agricultural commodities was the application of another turn to the screw of the "everyday" processes of accumulation and dispossession in rural areas. While one might have thought that higher food prices would be good for farmers, this was not generally the case, for several reasons. Prices rose in part because of higher costs for agricultural inputs such as oil and fertilizers,[4] and this did not benefit farmers at all. Large numbers of Southeast Asian farmers, and especially the poorest, are net food buyers rather than sellers, so they lost when prices rose; correspondingly, farmers with stocks of rice and with cash to make loans were positioned to accumulate. For all farmers, too, the volatility and unpredictability of prices made decision making in agriculture increasingly difficult.

The food crisis thus made life tougher for most farmers, but especially for the poorest, for whom it intensified tendencies towards dispossession. It did so, as well, at a time when opportunities for expelled farmers in the urban or industrial economies were shrinking rapidly, as export-oriented manufacturing collapsed. Indeed, one consequence of the crisis in the urban and industrial sectors was the reappearance of a discourse about the ability of the Southeast Asian countryside to serve as a "safety net" for redundant workers, who could return to "their" villages to wait out the crisis. The Asian Development Bank's country director for Thailand, for instance, argued that the Thai rural sector would be able to play this role.[5] State officials, and policymakers in international institutions such as the World Bank, also made this argument during the 1997–8 crisis; but, as we noted in Chapter 6, the "safety net" capabilities of the countryside were greatly exaggerated even then. In Java, many of the people who had left villages for the cities were precisely those who had lost their land, decades or sometimes generations previously, and they had also been squeezed out of opportunities to work as sharecroppers or labourers in the fiercely competitive agrarian scene we described. That the concept of the village as a safety net providing "farm-financed social welfare" re-emerged and was again promoted by the World Bank after another decade of deagrarianization, and at a time when agriculture in the region was facing a profound crisis, seems perverse.[6] More generally, the combination of new and intensified processes of exclusion we have examined in this book—forests zoned for conservation; land under oil palm; land covered by water or concrete; and village land more tightly mapped, titled and perhaps held by absentees—has reduced the ability of people who cannot find work in the city or countryside to provision themselves directly from the land. The quantitative extent of this problem, as an increased population experiences ever-more exclusionary conditions of land access, is historically unprecedented in Southeast Asia, and it remains quite unclear how the new pressures will play out.

A further consequence of the 2008–9 global economic crisis that is highly relevant to this book's concerns was its impact on the export crops already vulnerable to volatile prices, and cycles of boom and bust. The commodities boom was good for some crops but caused sharply reduced demand for others.

In Malaysia, the price of palm oil rose from 1,920 ringgit per metric ton in February 2007 to a peak of 3,681 in March 2008, and then fell back to 1,553 ringgit by December. In November 2008, Malaysia's deputy commodities minister warned that the price collapse was pushing smallholder oil palm producers to the point of bankruptcy. The major oil palm corporations, meanwhile, used the vulnerability of smallholders and smaller plantations as an opportunity to accumulate land.[7] Thai shrimp farmers had to cope with a substantial drop in export prices.[8] Coffee farmers in both Indonesia and Vietnam suffered as a result of a collapse in prices,[9] and rubber was hit hard by the collapse of global demand for automobiles.[10] Prices for these crops rebounded in 2009, but their effect on producers that lost their land during the crisis was permanent.

Perhaps the most direct connection between the crisis and the concerns of this book was the dramatic reappearance in 2008 of foreign direct investment (FDI) in agricultural land and production in the South, a process widely seen as a new "land-grab". The panic in agricultural commodity markets in early 2008 created an intensified sense among governments and corporations in food-importing countries, notably in the Middle East and East Asia, that international markets for staple foods could no longer be trusted. In response, these countries signed dozens of new agreements with governments in Africa, Eastern Europe and Southeast Asia to purchase or lease land for export-oriented food and biofuel production.[11] Every state in Southeast Asia took part in such agreements (many of which involved rice farming) in 2008–9. Some of these deals were quite small in scale, but others were imposingly large. Sime Darby, one of the world's biggest oil palm plantation companies, announced in June 2008 that it planned to begin growing rice on 6,900 hectares of land in Sarawak, and the Chinese telecommunications company ZTE Corp expressed its intention to grow cassava on 100,000 hectares in southern Laos. The Kuwaiti government signed a US\$486 million loan agreement with the Cambodian government to invest in irrigation infrastructure to grow rice on 50,000 hectares of land in the Stung Sen tributary catchment of the Tonle Sap Lake.[12] And in perhaps the most extraordinary of all these deals, Indonesia's special envoy to the Middle East announced in March 2009 that the country would devote at least 2 million hectares of farmland to joint ventures with Saudi companies, including a US\$4.3 billion, 500,000-hectare rice project in West Papua being developed by the Saudi Binladin Group.[13] This kind of foreign investment in land in Southeast Asia is hardly new, but the deals made in 2008 marked a dramatically higher level of interest.[14]

While announcements of deals such as this entered the news media, the amount of detailed information available about them was very small. News items generally referred to the governments and companies involved, the crops to be grown, and the amount of land involved. However, almost none of the media reports gave any specific information about exactly where the land was; whether it was in a contiguous piece; what activities were taking place on it at the time of signing; who was selling or leasing it, and on what terms; how it would be farmed, and by whom; and what would happen to the food once it was harvested. These questions are all, of course, of the first order of importance. While this lack of detail makes it difficult to give an in-depth analysis of the dynamics of the new "land-grab" in Southeast Asia, we can make a start on unpacking the dynamics of the process by looking at the powers involved.

The obvious power with which to begin the analysis is the market. As was the case in our chapters on boom crops and land conversion, the immediate trigger for this surge of interest in FDI in agricultural land was demand: in this case, demand for land on which to grow both food and tree crops. While this is indeed the right place to start, the market dynamics in play in this case are unusual, indeed ironic, in a number of respects. First, rapidly rising demand in the market for arable land in the South was being driven by the dysfunctionality of a different market, the international market for rice and other staple food crops. Middle Eastern and East Asian states and firms became interested in FDI in land only because high prices, and moves by some states in the region to prohibit exports, made importing states view the market for food as highly unreliable. They justified their overseas land acquisitions in terms of the need to bring the supply of food for their populations more directly under their own control. Second, the negotiation of these sales and leases generally took place under conditions that were a far cry from the "willing buyer, willing seller" model of the World Bank and other promoters of free markets in land. Rather, governments were deeply involved on both sides. This was necessarily the case, because there simply is no private market in any Southeast Asian state for parcels of farmland measured in the tens or hundreds of thousands of hectares. If you want this kind of land in the region, you have to deal with the state. Third, and again ironically, much of this investment was slated to be in rice. While Southeast Asian countries are some of the world's top rice exporters, rice agriculture in the region has been in crisis for decades, so why would rice attract so much foreign interest? Part of the explanation, of course, was that rice prices were very high at the time. It also seemed that foreign companies were planning to bring new technology and economies of scale to the sector, and if so, there would be no contradiction in this large-scale FDI, in the same way that there is no contradiction in European supermarkets pouring investment into Southeast Asia while small-scale, inefficient food retailers go under. However, and while we have in general argued against "crop essentialism" in this book, only in a few areas (such as the Muda plain in Malaysia, or parts of the coastal plain in Java) can rice can be effectively produced on a large scale under a centralized, estate-like productive regime. And these schemes do not have a monopoly on relevant technologies such as irrigation, mechanical ploughs and combine harvesters, which are widely used by smallholders. For the most part, rice is efficiently grown as a smallholder crop. This element of rice production both increases the importance of knowing under what conditions and by whom food would be grown on all this land, and also suggests that some of the schemes would not be competitive.

We see here again, then, a common theme of this book: prices are powerful, but they have their effects only in contexts that differ dramatically from ideological visions of free markets. This means that the other three powers of exclusion must also be considered. While details about these deals are scarce, it is safe to assume that regulation would play a critical role in finding the land promised in the contracts. Just as the land on which corporations establish oil palm plantations is generally received from the government through concessions and leases, so these notional rice, cassava and maize plantations would likely be carved out of the political forest in many cases. But even the political forest is not empty. When the people who live in it now protest its assignment to foreign plantations, force is likely to come into play and legitimacy will be hard to muster. Territory and belonging are, as we have emphasized throughout this book, deeply interconnected. The sale or lease of part of the national territory to foreign interests is almost always viewed with suspicion in Southeast Asia, and indeed in many parts of the region sales are tightly restricted. Recall that in 2008, when these deals were being signed and announced, there was a food crisis, making prospects for their local popularity doubly remote. In Madagascar, a government fell in 2009 partly as a result of popular discontent over a massive project of this kind.

At base, large-scale agricultural schemes for food or plantation crops depend upon a concept of empty land—massive expanses of forest or scrub just waiting for productive investment. But such land is hard to find. A news report from Laos conveyed the gap between the expectations, and in this case the contractual agreements, and the facts on the ground: "Plantations promised 20,000 hectares or 50,000 hectares can't find it when they survey. They might get 50 hectares here or there. But they want huge plantations. The dream was Brazil." Plantation companies also encountered sabotage, in the form of burned seedlings and equipment.[15] Land was hard to find not only because the locals protested, but also because of the increasingly dense array of zoning mechanisms we described in Chapters 2, 3 and 5: Land and Forest zones, conservation areas, and catchments for hydro dams. Nevertheless, huge areas of land in Laos and elsewhere in the region have been signed away to agribusiness corporations, and these deals cannot simply be dismissed. Almost certainly, they will not be realized in the way their proponents imagined, but they have the potential to create zones of overlapping jurisdiction, radically unsettling access to land not just for smallholders but also for actors with other agendas such as community-based management, electricity production or, in the case of West Papua, an ethno-territorial claim.

Crises such as the one in 2007–9 are not only transformative, setting a chain of events in motion. They are also revelatory, in the sense that they bring out very clearly a set of underlying pressures and tendencies that are present in "normal" times but more difficult to discern. Hence, they reveal something about future directions, even though the shape of the future cannot be precisely foretold. For this case, the weight of our preceding analysis allows us to wrap these tendencies into a headline list to watch out for: land access, squeezed; vulnerability to erratic prices, acute; rural livelihoods and opportunities for off-farm work, fragile; legitimacy of land regulation, in question; and—most striking of all—a renewed faith in agriculture as a vehicle for national growth, and as a route to well-being for tens of millions of rural Southeast Asians. We believe that these processes and tendencies are not unique to the region, though we leave the resonances of our analysis for others to explore.

Notes

1. Thanks to Rob Cramb for this point.
2. Berman 1981: 47.
3. Headey and Fan 2008.
4. See Bradsher and Martin 2008.
5. *International Herald Tribune*, 28 Jan. 2009.
6. Li 2009a.
7. *Straits Times*, 23 Nov. 2008; Hoh 2009.
8. Olimpo 2009.
9. *Jakarta Post*, 11 Oct. 2008; Deutsche Presse-Agentur, 13 Nov. 2008.
10. Li 2009a; Deutsche Presse-Agentur, 13 Nov. 2008; on coffee in Indonesia, see *Jakarta Post*, 11 Oct. 2008.
11. Grain 2008. In addition to governments and agricultural corporations, private investment funds are also acquiring farmland. See Henriques 2008.
12. khmerization.blogspot.com/2008/08/kuwait-loans-cambodia-546-mln-plans.html.
13. BBC News, 16 June 2008; Grain 2008; Reuters, 24 Mar. 2009.
14. See, for instance, Barney 2008.
15. *Guardian*, 22 Nov. 2008.

Appendix: Annals of Southeast Asian Land History

In this appendix, we provide brief histories of land relations and land use in the countries of Southeast Asia. Reasons of space have forced a certain level of anachronism on us, in the sense that we have read the existing states back into history, ignoring changes in sovereignty and borders. One exception to this is our treatment of current-day Malaysia, which we break up into discussions of Peninsular Malaysia, Sabah and Sarawak—but even this approach hides the very complex relations of sovereignty and rule that characterized the British colonial period on the Peninsula. We have grouped the histories geographically and historically: we begin with the former French colonies of Vietnam, Cambodia and Laos, then move on to Thailand (also on the mainland), Malaysia (which is partly on the mainland and partly insular), and the archipelagos of Indonesia and the Philippines. While these accounts are almost absurdly brief, we hope they will be of some use to readers new to the region.

Vietnam

Precolonial	Relatively high agricultural population densities, particularly in deltas and coastal strips. History of southward movement of Viet (Nam Tien) for land colonization. Mekong Delta colonized in quasi-military land settlements.
Colonial period (1890s–1953)	Significant area remains as communal land (one-fifth of cultivated area of Red River Delta in 1930s), with support of colonial state, which sees communal land as providing a safety valve and counter-vailing power to wealthy local interests. Colonial regime promotes plantation agriculture (especially rubber and coffee) and settlement of highlands by lowland

	Viet, and sedentarization of minorities, to alleviate population pressures in lowlands and "civilize" the highlands. Strengthening of landlord class in old delta areas. Large area of land alienated to French planters: 104,000 ha in Tonkin, 168,300 ha in Annam, 606,500 ha in Cochinchina. Emergence of significant landlessness in Red River and Mekong deltas. Frequent peasant unrest. Party garners support in Viet Minh-controlled areas after 1945 in part by cancellation of debt to moneylenders and landlords, rent reform, and allocation of land seized from more oppressive landlords. Viet Minh keeps "patriotic" landlords onside for united front against the French.
North Vietnam Post-independence socialism (1954–75)	1954–56 "excessive" land reform, violence against many better-off peasants as well as landlords, radical land seizure and land reform, partial loss of control by Party over land redistribution, settling of petty conflicts in name of class conflict based on land. Land totalling 810,000 ha redistributed by 1956. Post-1956 admission of mistakes and excesses, but increased production following redistribution. Collectivization of land from 1959, 90% complete in North by 1968. Collectivization both to capitalize agriculture and to prevent re-emergence of landlord class. About 1 million people resettled from Red River Delta to "New Economic Zones" in the northwest uplands during 1960s. Agriculture in northern Vietnam remains fully collectivized through to late 1970s. 5% of land allocated for household own-account production, contributes up to 40% of output.
South Vietnam Post-independence US-supported regime (1954–75)	Land inequality greater in South, and increasingly the basis of social differentiation and associated tensions in Mekong Delta. Viet Cong mobilization based in part on peasant grievances. Ngo Dinh Diem rent reform and landholding ceilings from 1950s, Nguyen Van Thieu conducts "pre-emptive" land reform from 1970 under "land to the tiller" slogan, involving redistribution of land to poorer farmers. Strategic

	hamlets programme of US-supported southern regime concentrates population into larger settlements.
Liberation, reunification and socialism (1975–80)	Rapid collectivization of land and draught animals in Mekong Delta, with much foot-dragging. Drastic fall in agricultural output. Large-scale clearing and settlement of land in the Central Highlands. Ultimately up to 6 million people move from lowland to upland areas through organized and spontaneous movement. Emphasis on provincial food self-sufficiency in rice.
Early reforms (1980s)	Contract system from 1980 under Decree 100 sees return to household-level production under contract to the cooperative, which provides agricultural services in exchange for fixed quota of rice, while land remains nominally collectively owned. De facto production increasingly household- rather than communally-oriented, and reform follows rather than leads peasant land tenure arrangements.
Doi Moi (Renovation) (1986–93)	*Doi Moi* period from Sixth Party Congress in 1986. Decree 10 in 1988 allows for allocation of land to households, marks more complete and formalized return to household economy. Cooperatives relegated to residual role in providing non-divisible services such as irrigation and plant protection.
Market-based growth (1993–)	Rise of land prices and establishment of de facto land market. 1993 Land Law allows land use certificates: 20-year leases on annually cropped land with a 3 ha ceiling, and 50-year leases on perennially planted land with a 10 ha ceiling in lowlands and 30 ha ceiling in uplands. Buying and selling of such leases permitted. Allocation of forestland to household level in uplands from 1994, encouraging forestry on "bare hills". Emerging land-based ethnic tensions in the Central Highlands with rapid expansion of coffee cultivation. Dispossession, including sale of land by ethnic minorities to newcomers. Loss of forest resources and increasing competition over groundwater exacerbates competition for land, as does

	restriction on shifting cultivation practices. Increased landlessness and indebtedness in the Mekong Delta. Land-based disputes between farmers and local authorities associated with development-related impacts such as road widening, golf courses, coastal resort development. Rapid conversion of mangroves to aquaculture, notably shrimp farms, providing another land frontier option for poor farmers in the southern Mekong Delta provinces.

Cambodia

Precolonial (Prior to 1863)	Land in principle the property of the King, in practice open for use. Labour and draught animals are the main constraint on developing and using paddy rice land: "acquisition by the plough". Abundant land relative to population. Most agricultural land is in a limited zone around the Tonle Sap Lake and along the Tonle Sap and Mekong rivers (a distribution that lasted until the late 20th century).
French colonial era (1863–1953)	Land Act 1884, not fully implemented until 1930s. Little by way of rural land registration. Cadastral system introduced 1912, mainly oriented to creating security for French investors. Civil Code of 1920 seen as moment of recognition of private property, including land.
Post-independence Sihanouk and Lon Nol period (1954–75)	Independence, but continuity in legal basis for land until 1975. 1956 Constitution specifically provides for private property rights, reiterated in 1972 Constitution. Limited progress in formal registration of land. Law applied only in more densely populated rice-growing areas. Landlessness steadily increases from 4% in 1950 to 20% in 1970. Massive dislocation of people from rural areas during early 1970s with US bombing and civil conflict; refugees from countryside abandon land and swell Phnom Penh population to 2 million (out of total 7 million).

Khmer Rouge (1975–78)	Khmer Rouge come to power under Pol Pot. Evacuation of urban population to the countryside. 1976 Constitution abolishes private property, including land. Radical collectivization, abolition of private property and money. Extensive "remodeling" of land in forced-labour construction of irrigation schemes along a grid pattern with little reference to micro-topography. Loss of 1 million–3 million lives. All cadastral records destroyed. Professional class wiped out or flees. Bunds and other boundary markers destroyed with collectivization.
Vietnamese Occupation— Heng Samrin regime (1979–91)	Break-up of Khmer Rouge order. Starvation and large-scale refugee exodus to Thailand and elsewhere. Civil war, with widespread mining of countryside. 30% decline in paddy land cultivation between 1967 and 1997. 1981 Constitution vests land as property of the state. Break-up of communes, although initially with continuation of small-scale collective farming ("solidarity groups", or *krom samakhi*). Repair or reconstruction of bunds, field boundaries for water management but not necessarily along old lines. Residential land use rights allocated on basis of occupation, land remains state owned in principle but individual peasant farming re-emerges in practice. 1989 Constitution reinstates private property, and 1989 Instruction #3 re-establishes private property and nullifies pre-1979 land claims. This allows possession of up to 5 ha for annual crops and concessions for plantation crops over 5 ha.
Coalition government (1991–97)	Paris Peace Agreements, emergence of stable order, decline in civil conflict. Khmer Rouge confined to more remote border areas in western and northern parts of the country. Return migration with repatriation of overseas refugees, and return to home provinces of internal refugees. Rural land allocation based on family size, not on prior occupation or claims. 1992 Land Law allows application for occupancy certificates, not full ownership in rural areas. System unable to cope

	with 4.5 million applications lodged; fewer than 14% are issued, mainly in accessible areas close to Phnom Penh and Siem Reap. Emergence of rural market economy, including market in land. Confusion as tax certificates, survey papers, application receipts used as documentation in such exchanges.
Entrenchment of Cambodian People's Party (1997–early 2000s)	Large-scale private interests associated with military and CPP leadership. Intensification of land conflict and land-grabbing. Rise in land values in remoter areas such as Ratanakiri (NE) with plantation crops, large foreign concessions (Indonesian, Malaysian) for oil palm, rubber. 2001 Land Law supersedes 1992 Act. Three domains recognized: state-public, state-private and private. Private domain includes individual, communal and shared land.
Economic growth and resource grabbing (2000s)	Land titling programme (Land Management and Administration Project, LMAP) from 2004 supported by World Bank, Finland and Germany, mainly in more accessible near-urban areas. Intensified land-grabbing. Boom in Siem Reap associated with growth of tourism at Angkor. Zoning of World Heritage Site places regulatory restriction on residential and agricultural land use. Growing landlessness, despite continuing overwhelming dependence on farming as main source of livelihood, and growing inequality. Poorly developed means for dispute resolution, continuing use of brute force. Backlog of court cases around land.

Laos

Precolonial	Very low population density, reduced even further by forced population movement to Siam from 1820s. Abundant land relative to population. Land in principle royal property but in practice held by peasant tillers. Land held by usufruct, with village-level arbitration.

French period (1893–1954)	No formalization of land tenure, no land register established. Unlike the other Indochina Union states, no tax on land or other property. Opium a major source of colonial revenue. French coffee plantations established on Boloven Plateau, but otherwise little colonial investment in rural land. Most of country engaged in subsistence-oriented farming.
Post-independence period (1954–75)	Land continues to be held by the King in principle and by farmers in practice up to 1975. No formalization of land rights or property. US bombing from the mid-1960s forces abandonment of land, even whole provinces in the case of Xieng Khouang (Plain of Jars), internal refugee movement, until by 1975 one-quarter of population internally displaced. Much cultivation is at survival, subsistence level.
Post-revolutionary period (1975–78)	Cooperatives established according to socialist model, but only in minority of villages. 4,000 cooperatives established between 1976 and 1986, but few remain functional as collectivized entities after the late 1970s. Collectivization rarely goes beyond mutual aid. Establishment of small numbers of state farms and resettled soldier settlements. Drastic decline in production under collective farming. Cooperatives short-lived and extended only to certain parts of the country.
Quasi-socialized agricultural economy (1980s)	Break-up of cooperatives. Land returned mainly to former owners. Mechanization limited to showcase state farms, insignificant in relation to total production area. Land remains basis of livelihood for 80–90% of the population. Provincial-level production targets aimed at rice self-sufficiency lead to expansion of hill rice farming, especially in northern provinces. Even schoolteachers and local government officers work land for basic subsistence (rice, kitchen gardens).
Market reforms (1986–96)	Market economy introduced following New Economic Mechanism of 1986, with gradual adoption of cash crops but continuing emphasis on food production with glutinous rice as staple. Tropical Forestry Action

	Plan and other international conservation groups encourage rationalization of land-based production, land use planning and "stabilization" of shifting cultivation. Decree 169 in 1993 provides for Land and Forest Allocation (LFA) at the village level, later termed *baeng din baeng paa* (dividing up land and forest) and then *mob din mob paa* (handing over land and forest). Land certificates issued for individual agricultural plots in more accessible areas, largely for tax purposes. Zoning under LFA gives each village a land and forest use plan displayed on a board at village entrance. Target of eradicating shifting cultivation by 2000 drives policy of resettlement and zoning.
Internationalized push for development (1997–2005)	AusAID/World Bank-funded land titling programme initiated in 1997, following the Thai model but working from urban and peri-urban areas outward. First phase to 2003 restricted to peri-urban Vientiane and a few provincial capitals. Second phase 2003–8 extends to rural areas, with debate about extent to which it should enter remoter mountainous areas. Priority given to areas that are market-linked and are economically more dynamic. AusAID cautious about extending too rapidly, and safeguard criteria established to avoid areas without reserved forest, areas with recent voluntary or involuntary immigration, and villages where ethnic minorities have communal land tenure under customary arrangements. In practice, titling largely restricted to residential land and farmland immediately contiguous to village settlements. Intensified conservation logic drives policy, i.e., focal village policy to attract people to move down from remote upland areas. Investment in subsidized pump irrigation to intensify food production in lowlands through double cropping. 2001 Participatory Poverty Assessment finds that restrictions on shifting cultivation are a major cause of poverty.
Intensified land pressures (2006–)	Intensive resource development, including land-hungry activities such as dams, plantations and mining. Viet-

<table>
<tr><td></td><td>namese companies in the south and Chinese companies in the north invest in rubber plantations on land previously used for shifting cultivation. Some smallholder engagement in plantation crops in the north in particular. Japanese eucalyptus plantations taken from "secondary forest" encroach on villagers' lands in central provinces. In 2007–8, rapid shift in discussion of land from vision of Laos as a land-abundant country to one in which land is scarce and conflict is emerging. National Land Management Authority (NLMA) established 2008 to deal with emergent land issues. Moratorium on new concessions leads to rapid granting of concessions prior to cut-off date. Land titling programme under Department of Lands moved from Ministry of Finance to NLMA. NLMA experiments with new land titling processes, also proposes land use planning ahead of titling—ideological differences with land titling programme sharpen on this issue and also on issuing title on land the state wishes to reserve for future development (e.g., roadside land, special economic zones).</td></tr>
</table>

Thailand

Pre-modern era (to mid-19th century)	Power and status in pre-modern Siam based on control over people, not land, and expressed through patron-clientage in quasi-feudal *naai-phrai* relations. This includes property rights in people (*phrai*), a form of slavery. Discursively, status expressed in terms of status in lands (*sakdina*, which also translates as quasi-feudalism in Thai), but in effect mediated through control over labour. Livelihood and production land-based, but land is abundant.
Post-Bowring expansion of rice economy (1850–92)	Royal land grants in central Thailand come with the opening up of commercial rice growing through canals for transport and irrigation, notably the Rangsit scheme of the Siam Canals and Lands Company. Land frontier largely lawless. Limited definition of property

	rights in land, but tax payment gives cultivators right to sell, use and pass on land to heirs. Land prices increase relative to labour in Central Plains. Titles given as early as 1882 in provinces surrounding Bangkok for land owned rather than land used. Continuous tax receipts of 10 years or more give right to exclude. Lack of central records, overlapping claims and conflicts. Property rights in people decline in importance, abolished in 1905, corvée abolished in favour of head tax by 1899.
Modernization reforms (1892–1940)	Agricultural development promoted through expansion rather than intensive land development as part of political project to bring outlying regions under centralized administrative control of Bangkok. 1892 Land Law partly clarifies security of tenure, but no central registry, overlapping claims, conflict in Rangsit area during 1890s. Torrens land titling system adopted in 1901, central registry and provincial land offices based on cadastral surveys. 637,001 titled holdings by 1910, of which 593,069 in Central Plains. Restrictions on sale of public land in 1929 to prevent land speculation.
Early development (1950s–60s)	1954 Land Code establishes current basis for private land ownership: occupation certificate (SK1 based on proof of pre-1954 occupation certified by local officials), reservation licence (NS2, granted on condition of use within 3 years), exploitation testimonial (NS3, based on proven use of at least 75% of reserved land on regular basis, mortgageable), full land certificate (NS4, fully transferable title). Slow rate of issue. Territorial and cash crop expansion following World Bank programme and start of five-year plans, expansion of agricultural area into upland forest margins. 1964 Forest Reserve Act places 45% of national territory under forest administration, far more extensive than natural forest. Logging and roads continue to open up new areas. Increasing numbers of people with insecure tenure in northeast Thailand and on the east and west

	fringes of the Central Plains. Of 9.79 million ha of farmland in late 1960s, 1.2 million had title (NS4), 1.79 had exploitation testimonial (NS3), 6.4 million had no documents.
Intensified agrarian tension (1970s)	Agrarian-based tension during democratic period following October 1973 uprising; mobilization over exploitative tenancy and landlessness, tenancy legislation limiting the more exploitative sharecropping. Assassination of farmers' leaders, especially in northern Thailand. Establishment of Agricultural Land Reform Office (ALRO) in 1975 ostensibly to redistribute land, in fact mainly formalizes settlement in forest reserve areas or promotes new settlement. End of easily available land frontier.
Economic boom and environmental concern (1980s)	Government tries to stop expansion into forest reserve areas, while ALRO programme extends state recognition into forest reserve areas previously untitled and, during CPT insurgency, out of bounds to the state. Royal Forest Department issues conditional title under STK (utilization certificate) programme—conditions on use and transfer. World Bank/AusAID land titling programme commences in 1984. Land price boom of the late 1980s starts in urban areas but rapidly spreads across Thailand, bringing speculative windfalls in coastal areas. Boom spreads outwards and sharpens titled/untitled land price differential.
Development and conservation (1990s–)	Intensification of conflict over marginal land in forest reserves, especially over eucalyptus plantations, dams. *Khor jor kor* programme threatens resettlement of up to 1 million families in name of "poverty alleviation", especially in Northeast Thailand. Move towards simplification of public and private land certification, with ALRO title on public land and NS4 full Torrens title on private land. Effort to establish communal tenure through community forest legislation from 1991, but fails to get through parliament. 1997 crisis partly based on the real estate-induced "bubble economy".

	Scenic value of land increasingly a factor in urban purchase of title. Stricter enforcement of protected area boundaries sets limits to further encroachment on public lands but also disenfranchises those settled in such areas prior to establishment of parks.

Malaysia

Peninsular Malaysia

Precolonial (to 19th century)	Debates in academic literature over nature of precolonial land tenure. Population density low. Variety in agricultural production, including relatively settled wet rice (*sawah*), fruit trees, shifting cultivation; different kinds of rights to each type of land. Land rights held by village headmen, chiefs or rajahs, who receive revenue, crops, corvée labour in a hierarchical social structure.
Colonial period (19th century to 1957)	Colonial history complex, with different regions coming under British control at different times and to different degrees; history of land issues also diverse. Plantations (especially pepper, coffee in 19th century, rubber boom in early 20th) expand rapidly under European and Chinese ownership, with labour drawn largely from India. Smallholders also shift to rubber. From 1890s officials try to use land policy to promote immigration and create/preserve a landowning peasantry, but rural Malays remain "peripatetic" into early 20th century. System of land reservation developed from 1910s to 1930s, under which Malay holdings may not be alienated to non-Malays. By decolonization, high levels of rural stratification, landlessness and tenancy.
Developmentalism (1957–80s)	Green revolution and substantial government involvement in agriculture. Federal Land Development Authority (FELDA) established by British in 1956 (in response to Communist insurgency) to resettle rural Malays from areas of land shortage to areas with available cultivable land, undertakes schemes

	covering 1 million ha by 1990s. Most FELDA land planted with oil palm, and plantations shift towards oil palm from rubber. Other agencies promote in situ development of existing agriculture. Share of smallholders in rubber and oil palm production increases. Extent of rice agriculture in decline from 1970s. Fragmentation of landholdings.
Neo-liberalism (1990s–)	Reorganization of FELDA away from settlement and towards approach focused on agribusiness and private sector; privatization. Extensive deagrarianization and idle agricultural land.

North Borneo/Sabah

North Borneo Company (1877–1941)	North Borneo Company acquires North Borneo from Sultan of Brunei in 1877, receives charter from Britain to administer territory in 1881, runs territory as commercial concern. Company focuses on developing plantation agriculture by providing land title under Western legal norms to private companies. Plantation expansion is based on Chinese, Javanese and native labour and dominated first by tobacco and then by rubber. Native claims to land meant to be settled (through reservation of land or cash payment) in areas of interest to foreigners, but little happens until 1910s; native claims often pushed aside in favour of plantations. Natives combine swidden, permanent farming, gardens and fruit trees, produce for subsistence and the market. Chinese migrants also involved in smallholder agriculture.
British Crown Colony (1946–63)	Government promotion of agricultural development through in situ and resettlement schemes with limited impact. Indonesia becomes main source of migrant labour. Government encourages sedentarization of shifting cultivators, hill rice production falls.
Malaysian statehood (1963–80s)	Logging boom from 1960s to 1990s. Further promotion by various parastatals of in situ and resettlement

	schemes, and of opening of new agricultural land. Oil palm and cocoa production expand, cocoa boom in 1970s and 1980s.
Oil palm (1990s–)	Logging, cocoa booms end. Parastatals shift to plantation model. Very rapid oil palm expansion, oil palm becomes dominant land use. Continuing decline of swidden. Deagrarianization.

Sarawak

Brooke Rajahs (1841–1941)	James Brooke made Rajah of Sarawak by Sultan of Brunei in 1842. Land Regulations of 1863 claim all unoccupied and "waste" land for the government, but state policy in practice has limited impact on land use/tenure. Brooke courts selectively enforce customary law. Brookes are wary of large-scale investment, and commercial agriculture is largely based on smallholding swidden farmers combining production of subsistence and cash crops (coffee, pepper, rubber). Rubber takes off around 1910. Immigrant Chinese farming in lowlands.
British Crown Colony (1946–63)	Land Code of 1958 divides land into Mixed Zone, Native Area, Reserved, and Interior Area; claims to native customary land can only be made based on occupation prior to 1958. State efforts to reserve forests, discourage shifting cultivation.
Malaysian statehood (1963–80s)	Government continues to discourage shifting cultivation, promotes expansion of crop production under various schemes involving resettlement, use of customary lands, or large estates. Rubber boom in 1960s, series of pepper booms/busts. Logging boom in 1970s and 1980s, protest against it in 1980s.
Oil palm (1990s–)	Logging boom ends, oil palm becomes dominant agricultural land use in 1990s. Smallholder agriculture increasingly dominated by pepper, rubber, oil palm; shifting cultivation and rice of decreasing importance. Rising outmigration from upland communities.

Indonesia

Precolonial/early colonial	Dutch East India Company present in the archipelago from around 1600, but seldom intervenes in land and production, extracting profits mainly through trade. Population generally sparse, but often concentrated in the uplands to escape piracy, slave raiding, imported diseases and malaria. In parts of Java early in the 19th century land individually held, with landlessness, sharecropping and land rental. Taxes collected on landowning households that incorporate landless people as farm servants and also employ roving bands of landless workers.
Colonial period (1800–1940)	Dutch Crown takes over the colony in 1800. During British interregnum (1812–16) Sir Thomas Stamford Raffles begins shifting the tax base from a personalistic apanage system to one based on village units. Colonial "cultivation system" conscripts land and labour for the forced production of sugar by smallholders in lowland Java (1840–70). Forced production of coffee imposed on smallholders in the uplands of Java, northern Sulawesi and Sumatra (1840–1920). In 1870, the Agrarian Law and associated Domain Declaration recognize native ownership of land under permanent cultivation, but claim uncultivated forest and "wasteland" as state domain, available for lease to foreigners for plantation development. Plantation sector expands rapidly in Sumatra, using indentured Javanese workers. Law forbids the sale or mortgage of native land, but land transfers and concentration proceed apace as village elites privileged by colonial rule accumulate land. By 1905, 40% of Java's population is landless. Land frontiers close, as the beginning of "scientific" forestry on Java leads to exclusion of smallholders and initiates a century of struggle still ongoing. Forest boundaries outside Java are seldom demarcated or patrolled, though conservation provides one of several rationales for colonial authorities to resettle swidden cultivators from hills to valleys.

War years (1941–45)	Japanese Occupation, independence struggle and bids for regional autonomy disrupt agricultural production.
Sukarno (1945–65)	Period of political activism that includes popular mobilizations to reclaim land from plantations. In 1960 the Basic Agrarian Law (BAL) passed under Sukarno establishes a system of land classification, limits the size of private holdings, provides for the recognition of customary land rights and raises expectations for land redistribution on Java. Land reclaiming, known as "unilateral action", organized by peasant unions and the Indonesian Communist Party targeting plantations with expired leases and landowners with land in excess of the legal limit.
New Order asserts control (1965–70)	New Order begins with massacre of about 500,000 villagers in Java and Bali on grounds of Communist affiliation. Land reform promised under BAL is abandoned. Popular mobilization for access to land is strongly suppressed. 1967 Forest Law declares about 75% of the nation's territory "forest", under the control of Forest Department. Forest Department issues timber concessions for extensive forestlands outside Java to cronies of General Suharto, the new president. Department does not follow the legal process for gazetting forest, which requires notification of villagers. Village protests suppressed.
New Order developmentalism (1970–90)	Green Revolution in rice production in Java increases productivity. Gains are captured mostly by landowners receiving state subsidies. In some parts of Java, up to 70% of the rural population is landless. Official transmigration programmes that move landless people from Java and Bali to less densely populated islands provoke protests focused on land seizure and the outnumbering of locals by migrants. There are also spontaneous migration streams. Official figures for "lifetime migration" (people not born in the province where they were enumerated in 1995) show that Java

	and Bali lost 3.5 million people, while Sumatra gained 2.5 million and Kalimantan 1 million migrants.
New Order wanes (1990s)	Outside Java, popular mobilization against state-backed appropriation of land for logging and plantations finds political room for manoeuvre through donor-driven environmental agendas stressing biodiversity, customary rights and indigenous knowledge. Many NGOs form around these issues. Agrarian reform movement in Java and Sumatra begins to reassemble, recommences its programme of occupying former plantation and state-claimed forestland, and presses for implementation of the reforms promised in the 1960 BAL. Main actors are NGOs, intellectuals and peasant unions. Suharto regime ends in 1998 amidst financial crisis, shift to democratic politics.
Popular mobilization and decentralization (1998–)	After hard lobbying by politicians, negotiation among activists, and mass rallies by Java-based peasant unions, parliament passes Decree on Agrarian Reform and Natural Resource Management (TAP IX), which attempts to meld the two streams of activism around land: the customary rights/environment stream outside Java, and the agrarian reform stream focused in Java and parts of Sumatra. To implement the principles enunciated in the decree, many separate pieces of legislation would need to be revised. Decentralization devolves control over forest administration to districts, causing massive scramble for timber. It also devolves land administration, making the status of many national land laws uncertain and increasing the difficulty of implementing the reforms promised in the parliamentary decree. District governments in search of revenue grant new plantation and mining concessions, popular protests over land confiscation continue.
Government announces land reform but fails to operationalize (late 2000s)	In 2006, government announces national programme of land reform to distribute 8.2 million ha to landless people, but location of the land and criteria for selection of beneficiaries are unclear. Following the historical pattern, wet rice remains concentrated in Java and

	export crops such as rubber, oil palm, coconut, pepper, tobacco, sugar cane and cloves in Sumatra, closely followed by Kalimantan. Export crop production dominated by smallholders operating independently or under contract. Population distribution remains uneven, with density per square km ranging from 11–69 in Kalimantan and 35–191 in Sulawesi and Sumatra to 559–1,033 in Java and Bali.

Philippines

Classical colonialism (16th to early 19th century)	Spanish colonialism predicated on isolation of countryside from foreigners and international influences. Colonial state takes almost no interest in areas beyond Manila, which are governed largely by the various Catholic religious orders. Some land accumulation takes place, notably on friar estates of central Luzon and in areas around Cebu, which grow rice, fruit and vegetables for urban consumption. Also some degree of accumulation by local *indio* elites in other parts of the region. Sharecropping extensive. Extremely limited connections between agricultural production and international trade, very little state interest in promoting agriculture as a source of revenue. Mindanao largely uncolonized.
Late Spanish colonialism (mid-19th century–1898)	Opening to international trade of Manila in 1834 and several key ports in other parts of the archipelago in 1855, in conjunction with entry of British and US capital, major stimuli to expansion of monocropping for export. Sugar, tobacco, hemp, coconuts become important crops. Growth of agriculture takes place largely through expansion of cultivated land, rather than intensification; much land controlled by Chinese-*indio* mestizos, who emerge as local elites controlling haciendas across much of the colony. Sharecropping the dominant organizational form of agricultural production. Formalization of land tenure remains extremely rudimentary, despite efforts at reform in

	1880s and 1890s. Frontier exists as an option, though not without cost, for smallholders who find themselves pressed too hard. Mindanao remains largely outside of Spanish control.
US colonialism (1899–1935/46)	US reaches accommodation with landed elite, incorporates them into colonial state as politicians with local electoral-economic power bases around the country. Access to US market under preferential tariffs very important to sugar. Friar estates purchased by state and sold off in 1903; land mostly bought by elites, or quickly ends up in their hands as former tenant purchasers fall back into tenancy. US state more committed to land titling as vehicle for economic development, but schemes mostly fail, also (like late Spanish schemes) used by elites to grab land. Rapid increase of socio-economic inequality during this period (carrying on from late Spanish period). Small, millenarian agrarian movements in 1910s and 1920s, larger-scale but short-lived Sakdalista movement, Sakdalista rebellion in 1935 quickly crushed.
War years (1941–44)	Resistance to Japanese Occupation centred in peasant guerrilla movements.
The republic (1946–72)	Republican period in which political power at national level shifts among coalitions of oligarchs rooted in landholding (though diversifying). Hukbalahap peasant rebellion (based in Luzon) 1946–54. New People's Army (armed wing of the Communist Party) forms in 1969, still active in 2000s. Violent response to agrarian movements from both state and landlords; landlords have their own armed gangs. Efforts at agrarian reform under various presidents largely stymied by landlord power in Congress. More than 1 million people migrate from Luzon and the central islands to Mindanao between World War II and 1960—some with state help, though most fairly spontaneously. Migration out of the Philippines (notably to California) begins in this period.

	Agricultural growth stagnant compared to other parts of Southeast Asia.
Marcos dictatorship (1972–86)	Ferdinand Marcos (elected president in 1965 and 1969) imposes martial law in 1972, sweeps aside Congress. Expansion of new(er) crops such as bananas, pineapples and farmed shrimp, with more involvement from agribusiness as opposed to the old landlord class. Growth in agricultural production remains slow compared to rest of region. Relatively effective land reform on rice and corn land from 1972.
The republic redux (1986–)	Popular/agrarian forces involved in People Power movement against Marcos impose land reform on a Congress composed largely of reinstated landlords. Comprehensive Agrarian Reform Program the most extensive land reform initiative in Philippine history and helps structure land relations in subsequent decades, though evaluations of its success are mixed. Extensive mobilization of agrarian sector, largely through NGOs, to some degree through Communist Party. Conversion of agricultural land to urban uses, industry and tourism emerges as major source of conflict. Indigenous Peoples' Rights Act of 1997 makes possible recognition of collective rights to ancestral domains through National Commission on Indigenous People. Continued very high rates of out-migration.

Bibliography

Abinales, Patricio N., *Making Mindanao: Cotabato and Davao in the Formation of the Philippine Nation-State*. Manila: Ateneo de Manila University Press, 2000.

Abinales, Patricio N. and Donna J. Amoroso, *State and Society in the Philippines*. Lanham, MD: Rowman & Littlefield, 2005.

Acciaioli, Greg, "Mobilizing against the 'Cruel Oil': Dilemmas of Organizing Resistance against Palm Oil Plantations in Central Kalimantan", in *Reflections on the Heart of Borneo*, ed. G.A. Persoon and M. Osseweijer. Wageningen: Tropenbos International, 2008.

Aditjondro, George J., "Large Dam Victims and Their Defenders: The Emergence of an Anti-Dam Movement in Indonesia", in *The Politics of Environment in Southeast Asia: Resources and Resistance*, ed. P. Hirsch and C. Warren. London and New York: Routledge, 1998.

Afiff, Suraya, Noer Fauzi, Gillian Hart, Lungisile Ntsebeza and Nancy Peluso, "Redefining Agrarian Power: Resurgent Agrarian Movements in West Java, Indonesia", 2004, http://www.escholarship.org/uc/item/7rf2p49g#page-1.

Afiff, Suraya and Celia Lowe, "Claiming Indigenous Community: Political Discourse and Natural Resource Rights in Indonesia", *Alternatives* 32 (2007): 73–97.

Agergaard, Jytte, Niels Fold and Katherine V. Gough, "Global-Local Interactions: Socio-economic and Spatial Dynamics in Vietnam's Coffee Frontier", *The Geographical Journal* 175, no. 2 (2009): 133–45.

Agrawal, Arun, *Environmentality: Technologies of Government and the Making of Subjects*. Durham: Duke University Press, 2005.

Agrawal, Arun and K.H. Redford, "Conservation and Displacement: An Overview", in *Protected Areas and Human Displacement: A Conservation Perspective*, ed. K.H. Redford and E. Fearn. New York: Wildlife Conservation Society, 2007.

Ahmed, Faris, "In Defence of Land and Livelihood: Coastal Communities and the Shrimp Industry in Asia". CUSO, Inter Pares, Sierra Club of Canada and the Consumers Association of Penang, 1997.

Akram-Lodhi, A. Haroon, "Vietnam's Agriculture: Processes of Rich Peasant Accumulation and Mechanisms of Social Differentiation", *Journal of Agrarian Change* 5, no. 1 (2005): 73–116.

Aksornkoae, Sanit, Wattana Sugunnasil and Suthawan Sathirathai, "Analytical Background of the Case Studies and Research Sites: Ecological, Historical and Social Perspectives", in *Shrimp Farming and Mangrove Loss in Thailand*, ed. E.B. Barbier and S. Sathirathai. Cheltenham, UK: Edward Elgar, 2004.

Aksornkoae, Sanit and Ruangrai Tokrisna, "Overview of Shrimp Farming and Mangrove Loss in Thailand", in *Shrimp Farming and Mangrove Loss in Thailand*, ed. E.B. Barbier and S. Sathirathai. Cheltenham, UK: Edward Elgar, 2004.

Aksornkoae, Sanit, Ruangrai Tokrisna, Wattana Sugunnasil and Suthawan Sathirathai, "The Importance of Mangroves: Ecological Perspectives and Socio-economic Values", in *Shrimp Farming and Mangrove Loss in Thailand*, ed. E.B. Barbier and S. Sathirathai. Cheltenham, UK: Edward Elgar, 2004.

Alexander, Jennifer and Paul Alexander, "Shared Poverty as Ideology: Agrarian Relationships in Colonial Java", *Man* 17, no. 4 (1982): 597–619.

__________, "Protecting Peasants from Capitalism: The Subordination of Javanese Traders by the Colonial State", *Comparative Studies in Society and History* 33, no. 2 (1991): 370–94.

Allen, John, *Lost Geographies of Power*. Oxford: Blackwell, 2003.

Anderson, Benedict, "Murder and Progress in Modern Siam", in *The Spectre of Comparisons: Nationalism, Southeast Asia and the World*, ed. B. Anderson. London and New York: Verso, 1998.

Asian Development Bank (ADB) and ActionAid Vietnam, "Participatory Poverty and Governance Assessment: Dak Lak Province", 2003.

Askew, Mark, "The Cultural Factor in Rural-Urban Fringe Transformation: Land, Livelihood, and Inheritance in Western Nonthaburi", in *Thailand's Rice Bowl: Perspectives on Agricultural and Social Change in the Chao Phraya Delta*, ed. F. Molle and T. Srijantr. Bangkok: White Lotus, 2003.

Bachriadi, Dianto and Mustofa Agung Sardjono, "Local Initiatives to Return Communities' Control over Forest Lands in Indonesia: Conversion or Occupation?" Paper presented at the Conference of the International Association for the Study of Common Property. Bali, Indonesia, 19–23 June 2006.

Baird, Ian G. and Bruce Shoemaker, "Unsettling Experiences: Internal Resettlement and International Aid Agencies in Laos", *Development and Change* 38, no. 5 (2007): 865–88.

Baker, Chris, "Thailand's Assembly of the Poor: Background, Drama, Reaction", *South East Asia Research* 8, no. 1 (2000): 5–29.

Bakker, Karen, "The Politics of Hydropower: Developing the Mekong", *Political Geography* 18 (1999): 209–32.

Bakker, Laurens, Gerben Nooteboom and Rosanne Rutten, "Localities of Value: Ambiguous Access to Land and Water in Southeast Asia", *Asian Journal of Social Science* 38 (2010): 167–71.

Barkin, J. Samuel and George E. Shambaugh, "Hypotheses on the International Politics of Common Pool Resources", in *Anarchy and the Environment: The International Relations of Common Pool Resources*, ed. J.S. Barkin and G.E. Shambaugh. Albany: State University of New York Press, 1999.

Barney, Keith, "China and the Production of Forestlands in Lao PDR: A Political Ecology of Transnational Enclosure", in *Taking Southeast Asia to Market: Commodities, Nature, and People in the Neoliberal Age*, ed. J. Nevins and N.L. Peluso. Ithaca: Cornell University Press, 2008.

BBC News, *Malaysia to Boost Rice Production*. 16 June 2008.

Bello, Walden, Herbert Docena, Marissa de Guzman and Marylou Malig, *The Anti-development State: The Political Economy of Permanent Crisis in the Philippines*. London: Zed Books, 2004.

Benda-Beckmann, Franz von and Keebet von Benda-Beckmann, "Recreating the Nagari: Decentralization in West Sumatra", Max Planck Institute for Social Anthropology Working Papers No. 31, 2001, http://www.eth.mpg.de/pubs/wps/pdf/mpi-eth-working-paper-0031.pdf.

Berman, Marshall, *All that Is Solid Melts into Air: The Experience of Modernity*. New York: Simon and Schuster, 1981.

Bernstein, Henry, "Agrarian Classes in Capitalist Development", in *Capitalism and Development*, ed. L. Sklair. London: Routledge, 1994.

Bissonnette, Jean-François, "Claiming Territories, Defending Livelihoods: The Struggle of Iban Communities in Sarawak", in *Agricultural Expansion in Southeast Asia: Borneo in the Eye of the Storm*, ed. R. De Koninck, S. Bernard and J.-F. Bissonnette. Singapore: NUS Press, 2011.

Boomgaard, Peter, "Forests and Forestry in Colonial Java, 1677–1942", in *Changing Tropical Forests: Historical Perspectives on Today's Challenges in Asia, Australasia and Oceania*, ed. J. Dargavel, K. Dixon and N. Semple. Canberra: Centre for Resource and Environmental Studies, Australian National University, 1988.

———, "The Javanese Village as a Cheshire Cat: The Java Debate against a European and Latin American Background", *Journal of Peasant Studies* 18, no. 2 (1991): 288–304.

Borras, Saturnino M. Jr, "State-Society Relations in Land Reform Implementation in the Philippines", *Development and Change* 32, no. 3 (2001): 545–75.

———, "The Philippine Land Reform in Comparative Perspective: Some Conceptual and Methodological Implications", *Journal of Agrarian Change* 6, no. 1 (2006a): 69–101.

———, "Redistributive Land Reform in 'Public' (Forest) Lands? Lessons from the Philippines and Their Implications for Land Reform Theory and Practice", *Progress in Development Studies* 6, no. 2 (2006b): 123–45.

Borras, Saturnino M. Jr and Jennifer C. Franco, "Contemporary Discourses and Contestations around Pro-Poor Land Policies and Land Governance", *Journal of Agrarian Change* 10, no. 1 (2010): 1–32.

Bradsher, Keith and Andrew Martin, "Shortages Threaten Farmers' Key Tool: Fertilizer", *The New York Times*, 30 Apr. 2008.

Brandon, Michael and Katrina Wells, "Planning for People and Parks: Design Dilemmas", *World Development* 20, no. 4 (1992): 557–70.

Breman, Jan, *The Village on Java and the Early-Colonial State*. Rotterdam: Comparative Asian Studies Program, 1980.

———, "Labour and Landlessness in South and South-East Asia", in *Disappearing Peasantries?* ed. D. Bryceson. London: Intermediate Technology Publications, 2000.

Breman, Jan and Gunawan Wiradi. *Good Times and Bad Times in Rural Java*. Leiden: KITLV, 2002.

Brenner, Robert, "Agrarian Class Structure and Economic Development in Pre-Industrial Europe", in *The Brenner Debate: Agrarian Class Structure and Economic Development in Pre-Industrial Europe*, ed. T.H. Alston and C.H.E. Philpin. Cambridge: Cambridge University Press, 1985.

Brockington, Dan, Jim Igoe and Kai Schmidt-Soltau, "Conservation, Human Rights and Poverty Reduction", *Conservation Biology* 20, no. 1 (2006): 250–2.

Brosius, J. Peter, Anna Lowenhaupt Tsing and Charles Zerner, eds., *Communities and Conservation: Histories and Politics of Community-Based Natural Resource Management*. Walnut Creek, CA: Altamira Press, 2005.

Brown, Elaine C., "Grounds at Stake in Ancestral Domains", in *Patterns of Power and Politics in the Philippines: Implications for Development*, ed. J.F. Eder and R.L. Youngblood. Tempe: Arizona State University, 1994.

Buckland, Helen, "The Oil for Ape Scandal: How Palm Oil Is Threatening the Orang-Utan". Friends of the Earth, 2005.

Burns, Anthony, "Thailand's 20 Year Program to Title Rural Land". Background paper prepared for *World Development Report 2005*. Washington, DC: World Bank, 2004.

Check, Erika, "Roots of Recovery", *Nature* 438 (2005).

Chhay, Sokong, "Customary Land Rights of Indigenous People and Their Violation in Ratanakiri Province, Cambodia: An Assessment of the Government's Response". Faculty of Graduate Studies, Mahidol University, Bangkok, 2004.

Chirapanda, S., "The Thai Land Reform Program, Bangkok, ALRO", http://www.seameo. org/vl/landreform/frame.htm, 1998.

Colchester, Marcus, "Indigenous Peoples and Communal Tenures in Asia", in *FAO Corporate Document Repository*, n.d., http://www.fao.org/docrep/oo7/y5407t/y5407t07. htm, accessed 14 June 2007.

Colchester, Marcus, Norman Jiwan, Andiko, Martua Sirait, Asep Yunan Firdaus, A. Surambo and Herbert Pane, *Promised Land: Palm Oil and Land Acquisition in Indonesia: Implications for Local Communities and Indigenous Peoples*. Moreton-in-Marsh, UK, and Bogor, Indonesia: Forest Peoples Programme, Perkumpulan Sawit Watch, HuMA and the World Agroforestry Centre, 2006.

Cole, C.L., "The Place of Golf in U.S. Imperialism", *Journal of Sport and Social Issues* 26, no. 4 (2002): 331–6.

Cooke, Fadzilah Majid, "Vulnerability, Control and Oil Palm in Sarawak: Globalization and a New Era?" *Development and Change* 33, no. 2 (2002): 189–211.

Cowherd, Robert and Eric J. Heikkila, "Orange County, Java: Hybridity, Social Dualism and an Imagined West", in *Southern California and the World*, ed. Eric J. Heikkila and Rafael Pizarro. Westport, CT: Praeger, 2002.

Cramb, Rob, *Land and Longhouse: Agrarian Transformation in the Uplands of Sarawak*. Copenhagen: NIAS Press, 2007a.

⸺, "Reinventing Dualism: Policy Narratives and Modes of Oil Palm Expansion in Sarawak". Unpublished manuscript, University of Queensland, 2007b.

⸺, "Agrarian Transitions in Sarawak: Intensification and Expansion Reconsidered", in *Agricultural Expansion in Southeast Asia: Borneo in the Eye of the Storm*, ed. R. De Koninck, S. Bernard and J.-F. Bissonnette. Singapore: NUS Press, 2011.

Cramb, Rob, Carol J. Pierce Colfer, Wolfram Dressler, Pinkaew Laungaramsri, Quang Trang Le, Elok Mulyoutami, Nancy Lee Peluso and Reed L. Wadley, "Swidden Transformations and Rural Livelihoods in Southeast Asia", *Human Ecology* 37 (2009): 323–46.

Cramb, R.A., T. Purcell and T.C.S. Ho, "Participatory Assessment of Rural Livelihoods in the Central Highlands of Vietnam", *Agricultural Systems* 81 (2004): 255–72.

Dalrymple, Kate, Jude Wallace and Ian Williamson, "Land Policy and Tenure in Southeast Asia 1995–2005". Paper presented at the Third Regional FIG Conference. Jakarta, Indonesia, 2004.

Davidson, Jamie S., *From Rebellion to Riots: Collective Violence on Indonesian Borneo*. Madison: University of Wisconsin Press, 2008.

De Angelis, Massimo, "Separating the Doing and the Deed: Capital and the Continuous Character of Enclosures", *Historical Materialism* 12, no. 2 (2004): 57–87.

De Koninck, Rodolphe, "The Peasantry as the Territorial Spearhead of the State in Southeast Asia: The Case of Vietnam", *Sojourn (Social Issues in Southeast Asia)* 11, no. 2 (1996): 231–58.

_______, "Southeast Asian Agriculture Post-1960: Economic and Territorial Expansion", in *Southeast Asia Transformed: A Geography of Change*, ed. C.L. Sien. Singapore: ISEAS, 2003.

_______, "On the Geopolitics of Land Colonization: Order and Disorder on the Frontiers of Vietnam and Indonesia", *Moussons* 9 (2006): 35–59.

De Koninck, Rodolphe, Stéphane Bernard and Jean-François Bissonnette, "Agricultural Expansion: Focusing on Borneo", in *Agricultural Expansion in Southeast Asia: Borneo in the Eye of the Storm*, ed. R. De Koninck, S. Bernard and J.-F. Bissonnette. Singapore: NUS Press, 2011.

De Koninck, Rodolphe and Steve Dery, "Agricultural Expansion as a Tool of Population Redistribution in Southeast Asia", *Journal of Southeast Asian Studies* 28, no. 1 (1997): 1–26.

de Soto, Hernando, *The Mystery of Capital: Why Capitalism Triumphs in the West and Fails Everywhere Else*. New York: Basic Books, 2000.

Degen, Peter and Nao Thuok, "Historical, Cultural and Legal Perspectives on the Fishing Lot System in Cambodia", in *Common Property in the Mekong: Issues of Sustainability and Subsistence*, ed. M. Ahmed and P. Hirsch. Penang: International Center for Living Aquatic Resources Management, 2000.

Deininger, Klaus, *Land Policies for Growth and Poverty Reduction*. Oxford: World Bank and Oxford University Press, 2003.

Delcore, Henry, "The Racial Distribution of Privilege in a Thai National Park", *Journal of Southeast Asian Studies* 38, no. 1 (2007): 83–105.

Deutsche Presse-Agentur, "Financial Crisis Hits Vietnam Farmers", 13 Nov. 2008.

D'haeze, D., J. Deckers, D. Raes, T.A. Phong and H.V. Loi, "Environmental and Socio-economic Impacts of Institutional Reforms on the Agricultural Sector of Vietnam: Land Suitability Assessment for Robusta Coffee in the Dak Gan Region", *Agriculture, Ecosystems and Environment* 105, no. 1–2 (2005): 59–76.

Dick, H.W. and P.J. Rimmer, "Beyond the Third World City: The New Urban Geography of South-East Asia", *Urban Studies* 35, no. 12 (1998): 2303–21.

Domoto, K., "Environmental Issues in Laos", in *Laos Dilemmas and Options: The Challenge of Economic Transition in the 1990s*, ed. Mya Than and J.L.H. Tan. Singapore: ISEAS, 1997.

Doolittle, Amity, *Property and Politics in Sabah, Malaysia: Native Struggles over Land Rights*. Seattle: University of Washington Press, 2005.

Doutriaux, Sylvie, Charles Geisler and Gerald Shively, "Competing for Coffee Space: Development-Induced Displacement in the Central Highlands of Vietnam", *Rural Sociology* 73, no. 4 (2008): 528–54.

Dove, Michael, "The Chayanov Slope in a Swidden Society", in *Chayanov, Peasants, and Economic Anthropology*, ed. E.P. Durrenberger. New York: Academic Press, 1984.

————, "Smallholder Rubber and Swidden Agriculture in Borneo: A Sustainable Adaptation to the Ecology and Economy of the Tropical Rainforest", *Economic Botany* 47, no. 2 (1993): 136–47.

Down to Earth, "Behind the Central Kalimantan Violence", *Down to Earth: International Campaign for Ecological Justice in Indonesia* 49 (2001a).

————, "Special Report on Transmigration", *Down to Earth: International Campaign for Ecological Justice in Indonesia* (2001b).

Dressler, Wolfram and Sarah Turner, "The Persistence of Social Differentiation in the Philippine Uplands", *Journal of Development Studies* 44, no. 10 (2008): 1450–73.

Ducourtieux, Olivier, Jean-Richard Laffort and Silinthone Sacklokham, "Land Policy and Farming Practices in Laos", *Development and Change* 36, no. 3 (2005): 499–526.

Eakin, Hallie, Alexandra Winkels and Jan Sendzimir, "Nested Vulnerability: Exploring Cross-Scale Linkages and Vulnerability Teleconnections in Mexican and Vietnamese Coffee Systems", *Environmental Science & Policy* 12 (2009): 398–412.

Ecologist, "Whose Common Future? Reclaiming the Commons". Gabriola Island, BC: New Society Publishers, 1993.

Economist, "The Other Oil Spill", 26 June 2010.

Elmhirst, Rebecca, "Space, Identity Politics and Resource Control in Indonesia's Transmigration Programme", *Political Geography* 18, no. 7 (1999): 813–35.

Elson, Robert, *The End of the Peasantry in Southeast Asia: A Social and Economic History of Peasant Livelihood*. London: Macmillan Press, 1997.

Fahn, James David, *A Land on Fire: The Environmental Consequences of Thailand's Boom*. Boulder, CO: Westview Press, 2003.

Fauzi, Noer, "Land Titles Do Not Equal Agrarian Reform", *Inside Indonesia* (Oct.–Dec. 2009).

Fay, C. and M. Sirait, "Reforming the Reformists in Post-Soeharto Indonesia", in *Which Way Forward? People, Forests and Policymaking in Indonesia, Resources for the Future*, ed. C.J.P. Colfer and I.A.P. Resosudarmo. Washington, DC: Resources for the Future, 2002.

Feder, G., T. Onchan and Y. Chalamwong, "Land Policies and Farm Performance in Thailand's Forest Reserve Areas", *Economic Development and Cultural Change* 36, no. 3 (Apr. 1988): 483–501.

Federation of Farmers' Unions (FSPI), "WTO: The Enemy of the Peasant! Federation of Indonesian Peasant Union (FSPI) Indonesia", in *Impact of the WTO on Peasants in Southeast Asia and East Asia*. La Via Campesina and Food First Information and Action Network (FIAN), n.d.

Feranil, Salvador H., "Stretching the 'Limits' of Redistributive Reform: Lessons and Evidence from the Philippines under Neoliberalism", in *Reclaiming the Land: The Resurgence of Rural Movements in Africa, Asia and Latin America*, ed. S. Moyo and P. Yeros. London and New York: Zed Books, 2005.

Ferguson, Bruce W. and Michael L. Hoffman, "Land Markets and Effect of Regulation on Formal-Sector Development in Urban Indonesia", *Review of Urban and Regional Development Studies* 5 (1993): 51–73.

Firman, Tommy, "Land Conversion and Urban Development in the Northern Region of West Java, Indonesia", *Urban Studies* 34, no. 7 (1997): 1027–46.

———, "Rural to Urban Land Conversion in Indonesia during Boom and Bust Periods", *Land Use Policy* 17, no. 1 (2000): 13–20.

———, "Urban Development in Indonesia, 1990–2001: From the Boom to the Early Reform Era through the Crisis", *Habitat International* 26 (2002): 229–49.

———, "New Town Development in Jakarta Metropolitan Region: A Perspective of Spatial Segregation", *Habitat International* 28 (2004): 349–68.

Fisher, Bob, "The Social and Economic Impacts of Post-Tsunami Reconstruction", *Mekong Update and Dialogue* 9, no. 2 (2006): 5.

Fisher, Robert, Stewart Maginnis, William Jackson, Edmund Barrow and Sally Jeanrenaud, *Poverty and Conservation: Landscapes, People and Power*. Gland, Switzerland, and Cambridge, UK: IUCN, World Conservation Union, 2005.

Flaherty, Mark and Peter Vandergeest, "'Low-Salt' Shrimp Aquaculture in Thailand: Goodbye Coastline, Hello Khon Kaen!" *Environmental Management* 22, no. 6 (1998): 817–30.

Flaherty, Mark, Peter Vandergeest and Paul Miller, "Rice Paddy or Shrimp Pond: Tough Decisions in Rural Thailand", *World Development* 27, no. 12 (1999): 2045–60.

Fold, Niels, "Oiling the Palms: Restructuring of Settlement Schemes in Malaysia and the New International Trade Regulations", *World Development* 28, no. 3 (2000): 473–86.

Fold, Niels and Tina Svan Hansen, "Oil Palm Expansion in Sarawak: Lessons Learned by a Latecomer?" in *Environment, Development and Change in Rural Asia-Pacific: Between Local and Global?* ed. J. Connell and E. Waddell. London and New York: Routledge, 2007.

Forsyth, Timothy J., "Tourism and Agricultural Development in Thailand", *Annals of Tourism Research* 22, no. 4 (1995): 877–900.

Forsyth, Timothy and Andrew Walker, *Forest Guardians, Forest Destroyers: The Politics of Environmental Knowledge in Northern Thailand*. Seattle and London: University of Washington Press, 2008.

Fox, Jefferson, "Understanding a Dynamic Landscape: Land Use, Land Cover and Resource Tenure in Northeastern Cambodia", in *Linking People, Place and Policy: A GIScience Approach*, ed. S.J. Walsh and K.A. Crews-Meyer. Dordrecht: Kluwer Academic Publishers, 2002.

Fujita, Yayoi, Kaisone Phengsopha, Thoumthone Vongvisouk and Sithong Thongmanivong, "Post-socialist Land Reform in Lao PDR and Its Impact on Community Land and Social Equity". Paper presented at the 11th Biennial Global Conference of the International Association for the Study of Common Property (IASCP). Bali, Indonesia, 2006.

Fukuyama, Francis, *The End of History and the Last Man*. New York: Free Press, 1992.

Ganjanapan, Anan, *Local Control of Land and Forest: Cultural Dimensions of Resource Management in Northern Thailand*. Chiang Mai: Chiang Mai University Faculty of Social Sciences RCSD Monograph Series, 2000.

Gaveau, David L.A., Matthew Linkie, Suyadi, Patrice Levang and Nigel Leader-Williams, "Three Decades of Deforestation in Southwest Sumatra: Effects of Coffee Prices, Law Enforcement and Rural Poverty", *Biological Conservation* 142 (2009): 597–605.

Geddes, William Robert, *Nine Dayak Nights*. London: Oxford University Press, 1961.

Geertz, Clifford, *Agricultural Involution: The Processes of Ecological Change in Indonesia*. Berkeley: University of California Press, 1963.

Geisler, Charles, "A New Kind of Trouble: Evictions in Eden", *International Social Science Journal* 55, no. 1 (2003): 69–78.

Giang Phan Trieu, "More Harm than Good: Land Control Policies in the Central Highlands, Vietnam: A Case Study of Lam Ha District, Lam Dong Province". Paper presented at the conference "Revisiting Agrarian Transformations in Southeast Asia: Empirical, Theoretical and Applied Perspectives". Chiang Mai, Thailand, 13–15 May 2010.

Giovannucci, Daniele, Bryan Lewin, Rob Swinkels and Panos Varangis, "The Socialist Republic of Vietnam: Coffee Sector Report". Washington, DC: The World Bank Agriculture and Rural Development Department, 2004.

Glassman, Jim, "Primitive Accumulation, Accumulation by Dispossession, Accumulation by 'Extra-Economic' Means", *Progress in Human Geography* 30, no. 5 (2006): 608–25.

Glassman, Jim and Chris Sneddon, "Chiang Mai and Khon Kaen as Growth Poles: Regional Industrial Development in Thailand and Its Implications for Urban Sustainability", *Annals of the American Academy of Political and Social Science* 590 (2003): 93–115.

Goldblum, Charles and Tai-Chee Wong, "Growth, Crisis and Spatial Change: A Study of Haphazard Urbanisation in Jakarta, Indonesia", *Land Use Policy* 17 (2000): 29–37.

Goldman, Michael, "Eco-Governmentality and Other Transnational Practices of a 'Green' World Bank", in *Liberation Ecologies: Environment, Development, Social Movements*, ed. R. Peet and M.J. Watts. London: Routledge, 2004.

————, *Imperial Nature: The World Bank and Struggles for Social Justice in the Age of Globalization*. New Haven and London: Yale University Press, 2005.

Grain, *Seized! The 2008 Land Grab for Food and Financial Security*, http://www.grain.org/go/landgrab, accessed 25 Oct. 2008.

Greenfield, Gerard, *Vietnam and the World Coffee Crisis: Local Coffee Riots in a Global Context*, Mar. 2002, http://www.focusweb.org/publications/2002/Vietnam-and-the-world-coffee-crisis.html, accessed 19 July 2007.

Guardian, "The Resentment Rises as Villagers are Stripped of Holdings and Livelihood", 22 Nov. 2008.

Ha Dang Thanh and Gerald Shively, "Coffee Boom, Coffee Bust and Smallholder Response in Vietnam's Central Highlands". Unpublished paper, 2004.

————, "Coffee Boom, Coffee Bust and Smallholder Response in Vietnam's Central Highlands", *Review of Development Economics* 12, no. 2 (2008): 312–26.

Hall, Derek, "The International Political Ecology of Industrial Shrimp Aquaculture and Industrial Plantation Forestry in Asia", *Journal of Southeast Asian Studies* 34, no. 2 (2003): 251–64.

__________, "Smallholders and the Spread of Capitalism in Rural Southeast Asia", *Asia Pacific Viewpoint* 45, no. 3 (2004a): 401–14.

__________, "Explaining the Diversity of Southeast Asian Shrimp Aquaculture", *Journal of Agrarian Change* 4, no. 3 (2004b): 315–35.

__________, "Rethinking Primitive Accumulation in Rural Southeast Asia". Paper presented at the Berkeley Workshop on Environmental Politics. Berkeley: University of California, 2010.

Hansen, Tina Svan, "Spatio-Temporal Aspects of Land Use and Land Cover Changes in the Niah Catchment, Sarawak, Malaysia", *Singapore Journal of Tropical Geography* 26, no. 2 (2005): 170–90.

Hardy, Andrew, "Strategies of Migration to Upland Areas in Contemporary Vietnam", *Asia Pacific Viewpoint* 41, no. 1 (2000): 23–34.

Hart, Gillian, *Power, Labour and Livelihood: Processes of Change in Rural Java*. Berkeley and Los Angeles: University of California Press, 1986.

__________, "Agrarian Change in the Context of State Patronage", in *Agrarian Transformations: Local Processes and the State in Southeast Asia*, ed. Gillian Hart, Andrew Turton and Ben White. Berkeley and Los Angeles: University of California Press, 1989.

Hart, Gillian, Andrew Turton and Ben White, eds., *Agrarian Transformations: Local Processes and the State in Southeast Asia*. Berkeley and Los Angeles: University of California Press, 1989.

Harvey, David, *The New Imperialism*. Oxford: Oxford University Press, 2003.

Hayami, Yujiro, "Ecology, History, and Development: A Perspective from Rural Southeast Asia", *World Bank Research Observer* 16, no. 2 (2001): 169–98.

Hayami, Yujiro, A.R. Quisumbing and Lourdes S. Adriano, *Toward an Alternative Land Reform Paradigm: A Philippine Perspective*. Manila: Ateneo de Manila University Press, 1990.

Headey, Derek and Shenggen Fan, "Anatomy of a Crisis: The Causes and Consequences of Surging Food Prices", *Agricultural Economics* 39 (Supplement) (2008): 375–91.

Henley, David, "Credit and Debt in Indonesian History: An Introduction", in *Credit and Debt in Indonesia, 860–1930*, ed. D. Henley and P. Boomgaard. Singapore: ISEAS Publishing, 2009.

Henriques, Diana B., "Food Is Gold, so Billions Invested in Farming", *The New York Times*, 5 June 2008.

Hirsch, Philip, *Development Dilemmas in Rural Thailand*. Singapore: Oxford University Press, 1990a.

__________, "Forests, Forest Reserve, and Forest Land in Thailand", *The Geographical Journal* 156, no. 2 (1990b): 166–74.

__________, "Environmental and Social Implications of Nam Theun Dam, Laos: Working Paper No. 5". Sydney: Economic and Regional Restructuring Research Unit, Departments of Economics and Geography, University of Sydney, 1991.

__________, *Political Economy of Environment in Thailand*. Manila: Journal of Contemporary Asia Publishers, 1993.

__________, "Community Forestry Revisited: Messages from the Periphery", in *Community Forestry at a Crossroads: Reflections and Future Directions in the Development of Community Forestry*, ed. M. Victor, C. Lang and J. Bornemeier. Bangkok: RECOFTC, 1998a.

———, "Dams, Resources and the Politics of the Environment in Mainland Southeast Asia", in *The Politics of Environment in Southeast Asia: Resources and Resistance*, ed. P. Hirsch and C. Warren. London and New York: Routledge, 1998b.

———, "Global Norms, Local Compliance and the Human Rights-Environment Nexus: A Case Study of the Nam Theun II Dam in Laos", in *Human Rights and the Environment: Conflicts and Norms in a Globalizing World*, ed. L. Zarsky. London: Earthscan, 2002.

———, "Revisiting Frontiers as Transitional Spaces in Thailand", *Geographical Journal* 175, no. 2 (2009): 124–32.

———, "The Changing Political Dynamics of Dam Building on the Mekong", *Water Alternatives* 3, no. 2 (2010): 312–23.

Hirsch, Philip, Khamla Phanvilay and Kaneungnit Tubtim, "Nam Ngum, Lao PDR: Community-Based Natural Resource Management and Conflicts over Watershed Resources", in *Cultivating Peace: Conflict and Collaboration in Natural Resource Management*, ed. D. Buckles. Ottawa: International Development Research Centre (IDRC), 1999.

Hirsch, Philip and Bach Tan Sinh, *Social and Environmental Implications of Resource Development in Vietnam: The Case of Hoa Bing Reservoir*. Sydney: Research Institute for Asia and the Pacific, 1992.

Hirsch, Philip and Andrew Wyatt, "Negotiating Local Livelihoods: Scales of Conflict in the Se San River Basin", *Asia Pacific Viewpoint* 45 (2004): 51–68.

Hirtz, Frank, "The Discourse that Silences: Beneficiaries' Ambivalence towards Redistributive Land Reform in the Philippines", *Development and Change* 29, no. 2 (1998): 247–75.

———, "It Takes Modern Means to Be Traditional: On Recognizing Indigenous Cultural Communities in the Philippines", *Development and Change* 34, no. 5 (2003): 887–914.

Hoh, Raymond, "Malaysia: Oilseeds and Products Annual 2009". United States Department of Agriculture, 2009.

Hughes, Helen, "The Pacific Is Viable!" in *Issue Analysis*. Sydney: Centre for Independent Studies, 2 Dec. 2004.

Hughes, Ross and Fiona Flintan, *Integrating Conservation and Development Experience: A Review and Bibliography of the ICDP Literature*. London: International Institute for Environment and Development, 2001.

Human Rights Watch, *Repression of Montagnards: Conflicts over Land and Religion in Vietnam's Central Highlands*. New York, 2002.

Hüsken, Frans, "Cycles of Commercialization and Accumulation in a Central Javanese Village", in *Agrarian Transformations: Local Processes and the State in Southeast Asia*, ed. Gillian Hart, Andrew Turton and Ben White. Berkeley and Los Angeles: University of California Press, 1989.

Hüsken, Frans, and Benjamin White, "Java: Social Differentiation, Food Production, and Agrarian Control", in *Agrarian Transformations: Local Processes and the State in Southeast Asia*, ed. Gillian Hart, Andrew Turton and Ben White. Berkeley and Los Angeles: University of California Press, 1989.

Hutchison, Jane, "Land Titling and Poverty Reduction in Southeast Asia: Realising Markets or Realising Rights?" *Australian Journal of International Affairs* 62, no. 3 (2008): 332–44.

Information Centre for Agricultural and Rural Development (ICARD) and Oxfam, "The Impact of the Global Coffee Trade on Dak Lak Province, Viet Nam: Analysis and Policy Recommendations". ICARD, Oxfam Great Britain and Oxfam Hong Kong, 2002.

International Centre for Environmental Management, "Regional Report on Protected Areas and Development: Review of Protected Areas and Development in the Lower Mekong River Region". Indooroopilly, Queensland, 2003.

International Crisis Group (ICG), "Communal Violence in Indonesia: Lessons from Kalimantan". Jakarta and Brussels, 2001.

———, "Indonesia: Managing Decentralization and Conflict". Jakarta and Brussels, 2003.

International Herald Tribune, "Unemployed Asians Return to Villages", 28 Jan. 2009.

Jakarta Post, "Coffee Farmers Worry as Price Plummets", 11 Oct. 2008.

Jamieson, Neil, Le Trong Cuc and A. Terry Rambo, *The Development Crisis in Vietnam's Mountains*. Hawaii: East-West Center Special Reports No. 6, 1998.

John, Ashish and Chea Phalla, "Community-Based Natural Resource Management and Decentralised Governance in Ratanakiri, Cambodia", in *Communities, Livelihoods and Natural Resources: Action Research and Policy Change in Asia*, ed. Stephen Tyler. Ottawa: ITDG Publishing, 2006.

Johnson, Craig A., "'Pink Gold' and the Politics of Land in Thailand: Market Penetration, Community Transformation and the Global Shrimp Industry". Paper presented at the annual meeting of the Rural Sociology Association. Albuquerque, NM, 2001.

Jomo, K.S., Y.T. Chang and K.J. Khoo, *Deforesting Malaysia: The Political Economy and Social Ecology of Agricultural Expansion and Commercial Logging*. London: Zed Books, 2004.

Jørgensen, Bent, "Development and 'The Other Within': The Culturalisation of the Political Economy of Poverty in the Northern Uplands of Viet Nam". PhD thesis, Peace and Development Research, Goteborg, 2006.

Kaewkuntee, Dararat, "Land Tenure, Land Conflicts and Post-Tsunami Relocation in Thailand", *Mekong Update and Dialogue* 9, no. 2 (2006): 2–4.

Kamnap, Phan and Sy Ramony, "Strengthening Local Voices to Inform National Policy: Community Forestry in Cambodia", in *Communities, Livelihoods and Natural Resources: Action Research and Policy Change in Asia*, ed. Stephen Tyler. Ottawa: ITDG Publishing, 2006.

Kelly, Philip F., "Everyday Urbanization: The Social Dynamics of Development in Manila's Extended Metropolitan Region", *International Journal of Urban and Regional Research* 23, no. 2 (1999): 283–303.

———, *Landscapes of Globalisation: Human Geographies of Economic Change in the Philippines*. London and New York: Routledge, 2000.

———, "Urbanization and the Politics of Land in the Manila Region", *The Annals of the American Academy of Political and Social Science* 590 (2003): 170–87.

Kerkvliet, Benedict J., *The Huk Rebellion: A Study of Peasant Revolt in the Philippines*. Berkeley: University of California Press, 1977.

_______, "Claiming the Land: Take-overs by Villagers in the Philippines with Comparisons to Indonesia, Peru, Portugal and Russia", *Journal of Peasant Studies* 20, no. 3 (1993): 459–93.

_______, *The Power of Everyday Politics: How Vietnamese Peasants Transformed National Policy*. Ithaca: Cornell University Press, 2005.

_______, "Agricultural Land in Vietnam: Markets Tempered by Family, Community and Socialist Practices", *Journal of Agrarian Change* 6, no. 3 (2006): 285–305.

Klinken, Gerry van, *Communal Violence and Democratization in Indonesia: Small Town Wars*. London: Routledge, 2007.

_______, "Blood, Timber, and the State in West Kalimantan, Indonesia", *Asia Pacific Viewpoint* 49, no. 1 (2008): 35–47.

Kompas, "Pengungsi Aceh: Nyawa Bisa Dibeli?" 29 July 1999.

_______, "Warga non-Aceh Mulai Eksodus", 18 Nov. 1999.

_______, "Eks Transmigran Aceh Bingung, tak punya apa pun lagi", 20 Nov. 1999.

_______, "Salam buat Pejabat di Gedung Sate", 16 Oct. 2000a.

_______, "Transmigrasi Local di Jabar Selatan Menebar Harapan, Menjinakkan Bom Waktu", 16 Oct. 2000b.

_______, "Proyek Transmigrasi untuk Pengungsi Aceh", 24 Oct. 2000.

Kratoska, Paul H., "The Peripatetic Peasant and Land Tenure in British Malaya", *Journal of Southeast Asian Studies* 16, no. 1 (1985): 16–46.

Kuhonta, Eric Martinez, "Development and Its Discontents: The Case of the Pak Mun Dam in Northeastern Thailand", in *Agrarian Angst and Rural Resistance in Contemporary Southeast Asia*, ed. D. Caouette and S. Turner. London and New York: Routledge, 2009.

Labbé, Danielle, "Urban Expansion and Emerging Land Disputes in Periurban Hanoi during the Late-Socialist Period: The Case of Hoa Muc Village". Paper presented at the conference "Revisiting Agrarian Transformations in Southeast Asia: Empirical, Theoretical and Applied Perspectives". Chiang Mai, Thailand, 13–15 May 2010.

Lange, Andreas, "Elites and Local Development in the Philippines", *Development and Change* 41, no. 1 (2010): 53–76.

Lapera, Tim, ed., *Prinsip-prinsip Reforma Agraria: Jalan Penghidupan dan Kemakmuran Rakyat*. Yogyakarta: Lapera Pustaka Utama, 2001.

La Via Campesina and Food First Information and Action Network (FIAN), "Global Campaign for Agrarian Reform Working Document: Commentary on Land and Rural Development Policies of the World Bank", http://viacampesina.org/main_en/images/stories/pdf/Global_Campain_WB_policies_factsheet.en.pdf, n.d.

Leaf, Michael, "The Suburbanization of Jakarta: A Concurrence of Economics and Ideology", *Third World Planning Review* 16, no. 4 (1994): 341–56.

_______, "A Tale of Two Villages: Globalization and Peri-urban Change in China and Vietnam", *Cities* 19, no. 1 (2002): 23–31.

Lebel, Louis, Po Garden and Masao Imamura, "The Politics of Scale, Position, and Place in the Governance of Water Resources in the Mekong Region", *Ecology and Society* 10, no. 2 (2005).

Leonard, Rebeca and Kingkorn Narintrakul na Ayutthaya, "Taking Land from the Poor, Giving Land to the Rich", *Watershed* 8, no. 2 (2003): 14–37.

Lewis, Judith, "The Shrimp Factor: Did Disappearing Mangrove Forests Contribute to the Tsunami's Severity?" *LA Weekly*, 7–13 Jan. 2005.

Li, Tania Murray, "Images of Community: Discourse and Strategy in Property Relations", *Development and Change* 27, no. 3 (1996): 501–27.

———, "Articulating Indigenous Identity in Indonesia: Resource Politics and the Tribal Slot", *Comparative Studies in Society and History* 42, no. 1 (2000): 149–79.

———, "Masyarakat Adat, Difference, and the Limits of Recognition in Indonesia's Forest Zone", *Modern Asian Studies* 35, no. 3 (2001a): 645–76.

———, "Relational Histories and the Production of Difference on Sulawesi's Upland Frontier", *Journal of Asian Studies* 60, no. 1 (2001b): 41–66.

———, "Engaging Simplifications: Community-Based Resource Management, Market Processes and State Agendas in Upland Southeast Asia", *World Development* 30, no. 2 (2002a): 265–83.

———, "Ethnic Cleansing, Recursive Knowledge, and the Dilemmas of Sedentarism", *International Journal of Social Science* 173 (2002b).

———, "Local Histories, Global Markets: Cocoa and Class in Upland Sulawesi", *Development and Change* 33, no. 3 (2002c): 415–37.

———, "Adat in Central Sulawesi: Contemporary Deployments", in *The Revival of Tradition in Indonesian Politics: The Deployment of Adat from Colonialism to Indigenism*, ed. J.S. Davidson and D. Henley. London: Routledge, 2007a.

———, "Practices of Assemblage and Community Forest Management", *Economy and Society* 36, no. 2 (2007b): 263–93.

———, *The Will to Improve: Governmentality, Development, and the Practice of Politics*. Durham: Duke University Press, 2007c.

———, "Exit from Agriculture: A Step Forward or a Step Backward for the Rural Poor?" *Journal of Peasant Studies* 36, no. 3 (2009a): 629–36.

———, "To Make Live or Let Die? Rural Dispossession and the Protection of Surplus Populations", *Antipode* 41 (S1) (2009b): 66–93.

———, "Agrarian Class Formation in Upland Sulawesi, 1990–2010". Montreal: ChATSEA Working Paper, 2010a.

———, "Indigeneity, Capitalism, and the Management of Dispossession", *Current Anthropology* 51, no. 3 (2010b): 385–414.

Lohmann, Larry, *Ecology as Racism: Forest Cleansing Racial Oppression in Scientific Nature Conservation*. Sturminster Newton: Cornerhouse, 1999.

———, *Polanyi along the Mekong: New Tensions and Resolutions over Land*. Sturminster Newton: Cornerhouse, 2002.

Lucas, Anton and Carol Warren, "The State, the People, and Their Mediators: The Struggle over Agrarian Law Reform in Post-New Order Indonesia", *Indonesia* 76 (2003): 87–126.

Lynch, Owen J. and Kirk Talbott, *Balancing Acts: Community-Based Forest Management and National Law in Asia and the Pacific*. Washington: World Resources Institute, 1995.

Malaque, Isidoro R. III and Makoto Yokohari, "Urbanization Process and the Changing Agricultural Landscape Pattern in the Urban Fringe of Metro Manila, Philippines", *Environment and Urbanization* 19, no. 1 (2007): 191–206.

Maneepong, Chuthatip and Douglas Webster, "Governance Responses to Emerging Peri-urbanisation Issues at the Global-Local Nexus", *International Development Planning Review* 30, no. 2 (2008): 133–54.

Mann, Michael, *Sources of Social Power, Volume 1: A History of Power from the Beginning to AD 1760*. Cambridge: Cambridge University Press, 1986.

Marsh, Sally P. and T. Gordon MacAulay, "Land Reform and the Development of Commercial Agriculture in Vietnam: Policy and Issues". Department of Agricultural Economics, University of Sydney, 2006.

McAndrew, John P., *Urban Usurpation: From Friar Estates to Industrial Estates in a Philippine Hinterland*. Quezon City: Ateneo de Manila University Press, 1994.

McAuslan, Patrick, *Bringing the Law Back In: Essays in Land, Law and Development*. London: Ashgate, 2003.

McCarthy, John F., "The Changing Regime: Forest Property and Reformasi in Indonesia", *Development and Change* 31, no. 1 (2000): 91.

_______, "Changing to Gray: Decentralization and the Emergence of Volatile Socio-legal Configurations in Central Kalimantan, Indonesia", *World Development* 32, no. 7 (2004): 1199–223.

_______, *The Fourth Circle: A Political Ecology of Sumatra's Rainforest Frontier*. Stanford: Stanford University Press, 2006.

McCarthy, John F. and R.A. Cramb, "Policy Narratives, Landholder Engagement, and Oil Palm Expansion on the Malaysian and Indonesian Frontiers", *The Geographical Journal* 175, no. 2 (2009): 112–23.

McCarthy, John and Zahari Zen, "Regulating the Oil Palm Boom: Assessing the Effectiveness of Environmental Governance Approaches to Agro-Industrial Pollution in Indonesia", *Law & Policy* 32, no. 1 (2010): 153–79.

McDermott, Melanie Hughes, "Invoking Community: Indigenous People and Ancestral Domain in Palawan, the Philippines", in *Communities and the Environment: Ethnicity, Gender, and the State in Community-Based Conservation*, ed. A. Agrawal and C. Gibson. New Brunswick: Rutgers University Press, 2001.

McGee, Terry, "Urbanisasi or Kotadesasi? Evolving Patterns of Urbanization in Asia", in *Urbanization in Asia: Spatial Dimensions and Policy Issues*, ed. F.J. Costa, A.K. Dutt, L.J.C. Ma and A.G. Noble. Honolulu: University of Hawaii Press, 1989.

_______, "The Emergence of Desakota Regions in Asia: Expanding a Hypothesis", in *The Extended Metropolis: Settlement Transition in Asia*, ed. N. Ginsburg, B. Koppel and T. McGee. Honolulu: University of Hawaii Press, 1991.

McKenny, Bruce and Prom Tola, "Natural Resources and Rural Livelihoods in Cambodia", CDRI Working Paper 23. Phnom Penh: Cambodia Development Resource Institute, 2002.

McMorrow, Julia and Mustapa Talip, "Decline of Forest Area in Sabah, Malaysia: Relationship to State Policies, Land Code and Land Capability", *Global Environmental Change* 11, no. 3 (2001): 217–30.

Mekhora, Thamrong and Laura M.J. McCann, "Rice versus Shrimp Production in Thailand: Is There Really a Conflict?" *Journal of Agricultural and Applied Economics* 35, no. 1 (2003): 143–57.

Menasveta, Piamsak, "Mangrove Destruction and Shrimp Culture Systems", *World Aquaculture* 28, no. 4 (1997): 36–42.

Missingham, Bruce, *The Assembly of the Poor in Thailand: From Local Struggles to Nation Protest Movement.* Chiang Mai: Silkworm Books, 2003.

Mitchell, Michael, "The Political Economy of Mekong Basin Development", in *The Politics of Environment in Southeast Asia: Resources and Resistance*, ed. P. Hirsch and C. Warren. London and New York: Routledge, 1998.

Miyamoto, Motoe, "Forest Conversion to Rubber around Sumatran Villages in Indonesia: Comparing the Impacts of Road Construction, Transmigration Projects and Population", *Forest Policy and Economics* 9 (2006): 1–12.

Morales-Fernholz, Rosemary, "Indigenous Land Rights: Who Controls the Philippine Public Domain?" in *Sovereignty under Challenge: How Governments Respond*, ed. J.D. Montgomery and N. Glazer. New Brunswick, NJ: Transaction Publishers, 2002.

Murakami, Akinobu, Alinda Medrial Zain, Kazuhiko Takeuchi, Atsushi Tsunekawa and Shigehiro Yokota, "Trends in Urbanization and Patterns of Land Use in the Asian Mega Cities Jakarta, Bangkok, and Metro Manila", *Landscape and Urban Planning* 70 (2005): 251–9.

Nation, "Bring 'Dead Assets' to Life, de Soto Advises Govt.", 8 Nov. 2002.

Ngidang, Dimbab, "Contradictions in Land Development Schemes: The Case of Joint Ventures in Sarawak, Malaysia", *Asia Pacific Viewpoint* 43, no. 2 (2002): 157–80.

Nguyen, T.V., "Country Report: Vietnam", Asia-Pacific Forestry Sector Outlook Study, Working Paper No. APFSOS/WP/31. Forestry Policy and Planning Division, Rome, and Regional Office for Asia and the Pacific, Bangkok, 1997.

Nguyen Van Suu, "Contending Views and Conflicts over Land in Vietnam's Red River Delta", *Journal of Southeast Asian Studies* 38, no. 2 (2007): 309–34.

Nong, Kim and Melissa Marschke, "Building Networks of Support for Community-Based Coastal Resource Management in Cambodia", in *Communities, Livelihoods and Natural Resources: Action Research and Policy Change in Asia*, ed. Stephen Tyler. Ottawa: ITDG Publishing, 2006.

O'Brien, Kevin, "Rightful Resistance", *World Politics* 49, no. 1 (1996): 31–55.

Olimpo, "Thai Aquaculture Seeks Opportunity amid Crisis", *FEED Business Asia*, Mar./Apr. 2009.

Parsons, Kenneth, "The Institutional Basis of an Agricultural Market Economy", *Journal of Economic Issues* 8, no. 4 (1974): 737–57.

Peet, Richard and Michael Watts, "Introduction: Development Theory and Environment in an Age of Market Triumphalism", *Economic Geography* 69, no. 3 and 4 (1993): 227–53.

Peluso, Nancy Lee, "A Political Ecology of Violence and Territory in West Kalimantan", *Asia Pacific Viewpoint* 49, no. 1 (2008): 48–67.

Peluso, Nancy Lee, Suraya Afiff and Noer Fauzi Rachman, "Claiming the Grounds for Reform: Agrarian and Environmental Movements in Indonesia", *Journal of Agrarian Change* 8, no. 2 and 3 (2008): 377–407.

Peluso, Nancy Lee and Emily Harwell, "Territory, Custom, and the Cultural Politics of Ethnic War in West Kalimantan, Indonesia", in *Violent Environments*, ed. Nancy Lee Peluso and Michael Watts. Ithaca: Cornell University Press, 2001.

Peluso, Nancy Lee and Peter Vandergeest, "Genealogies of the Political Forest and Customary Rights in Indonesia, Malaysia, and Thailand", *Journal of Asian Studies* 60, no. 3 (2001): 761–812.

Perelman, Michael, *The Invention of Capitalism: Classical Political Economy and the Secret History of Primitive Accumulation*. Durham and London: Duke University Press, 2000.

Perez, Padmapani and Tessa Minter, "Indigenous Rights and Resource Management in Philippine Protected Areas", *IIAS Newsletter* 35 (2004).

Phanvilay, Khamla, Kaneungnit Tubtim and Philip Hirsch, "Decentralisation, Watersheds and Ethnicity in Laos", in *Resources, Nations and Indigenous Peoples: Case Studies from Australia, Melanesia and Southeast Asia*, ed. R. Howitt, J. Connell and P. Hirsch. Melbourne: Oxford University Press, 1996.

Phnom Penh Post, "Govt's Land Policy Failing Most Vulnerable: Report", 28 Sept. 2009.

Phongpaichit, Pasuk and Chris Baker, *Thailand: Economy and Politics*. Oxford: Oxford University Press, 1995.

__________, *Thailand's Boom and Bust*. Chiang Mai: Silkworm Books, 1998.

Pincus, Jonathan, *Class Power and Agrarian Change: Land and Labour in Rural West Java*. London: Macmillan Press, 1996.

Polanyi, Karl, *The Great Transformation: The Political and Economic Origins of Our Time*. Boston: Beacon Press, 1957.

Popkin, Samuel L., *The Rational Peasant*. Berkeley: University of California Press, 1979.

Potter, Lesley, "The Oil Palm Question in Borneo", in *Reflections on the Heart of Borneo*, ed. G.A. Persoon and M. Osseweijer. Wageningen: Tropenbos International, 2008.

__________, "Oil Palm and Resistance in West Kalimantan, Indonesia", in *Agrarian Angst and Rural Resistance in Contemporary Southeast Asia*, ed. D. Caouette and S. Turner. London and New York: Routledge, 2009.

__________, "Kalimantan in the Firing Line: A Note on the Effects of the Global Financial Crisis", *Bulletin of Indonesian Economic Studies* 46, no. 1 (2010): 99–109.

Rattanabirabongse, V., R.A. Eddington, A.F. Burns and K.G. Nettle, "The Thailand Land Titling Project: Thirteen Years of Experience", *Land Use Policy* 15, no. 1 (1998): 3–23.

Redford, Kent and Eva Fearn, eds., *Protected Areas and Human Displacement: A Conservation Perspective*. New York: Wildlife Conservation Society, 2007.

Reid, Anthony, ed., *Slavery, Bondage, and Dependency in Southeast Asia*. New York: St Martin's Press, 1983.

Reuters, "Saudis Invest $1.3 Billion in Indonesian Agriculture", 24 Mar. 2009.

Ribot, Jesse C. and Nancy Lee Peluso, "A Theory of Access", *Rural Sociology* 68, no. 2 (2003): 153–81.

Rigg, Jonathan, "Thailand's Nam Choan Dam Project: A Case Study in the 'Greening' of South-east Asia", *Global Ecology and Biogeography* 1 (1991): 42–54.

__________, *More than the Soil: Rural Change in Southeast Asia*. London: Prentice Hall, 2001.

__________, "Land, Farming, Livelihoods, and Poverty: Rethinking the Links in the Rural South", *World Development* 34, no. 1 (2006): 180–202.

Rigg, Jonathan and Albert M. Salamanca, "Tensions between AFPS Production Systems and Other Land Uses". Production in Aquatic Peri-Urban Systems in Southeast Asia (PAPUSSA), 2006.

Ruf, François and Yoddang, "Cocoa Migrants from Boom to Bust", in *Agriculture in Crisis: People, Commodities and Natural Resources in Indonesia, 1996–2000*, ed. F. Gérard and F. Ruf. Montpellier, France: Cirad, 2001.

Ruf, François, Yoddang and Waris Ardhy, "Transmigrants and the Cocoa Windfall: 'Paradise Is Here, Not in Bali'", in *Agriculture in Crisis: People, Commodities and Natural Resources in Indonesia, 1996–2000*, ed. F. Gérard and F. Ruf. Montpellier, France: Cirad, 2001.

Rusanen, Liisa, "In Whose Interests? The Politics of Land Titling". Background paper. Sydney: AID/WATCH, 2005.

Rutten, Rosanne, "Who Shall Benefit? Conflicts among the Landless Poor in a Philippine Agrarian Reform Programme", *Asian Journal of Social Science* 38 (2010): 204–19.

Sajor, Edsel E., "Globalization and the Urban Property Boom in Metro Cebu, Philippines", *Development and Change* 34, no. 3 (2003).

Sangaji, Arianto, "The Masyarakat Adat Movement in Indonesia: A Critical Insider's View", in *The Revival of Tradition in Indonesian Politics: The Deployment of Adat from Colonialism to Indigenism*, ed. J.S. Davidson and D. Henley. London: Routledge, 2007.

Scott, James C., *The Moral Economy of the Peasant*. New Haven: Yale University Press, 1976.

———, *Weapons of the Weak: Everyday Forms of Peasant Resistance*. New Haven: Yale University Press, 1985.

———, *Seeing Like a State: How Certain Schemes to Improve the Human Condition Have Failed*. New Haven: Yale University Press, 1998.

———, *The Art of Not Being Governed: An Anarchist History of Upland Southeast Asia*. New Haven: Yale University Press, 2009.

Seng, Sovathana, "The Transformation of Northeastern Cambodia: The Politics of Development in an Ethnic Minority Community, Ratanakiri Province". MA thesis in Sustainable Development, Chiang Mai University, 2004.

Shatkin, Gavin, "Obstacles to Empowerment: Local Politics and Civil Society in Metropolitan Manila, the Philippines", *Urban Studies* 37, no. 12 (2000): 2357–75.

———, "Globalization and Local Leadership: Growth, Power and Politics in Thailand's Eastern Seaboard", *International Journal of Urban and Regional Research* 28, no. 1 (2004): 11–26.

———, "The City and the Bottom Line: Urban Megaprojects and the Privatization of Planning in Southeast Asia", *Environment and Planning A* 40, no. 2 (2008): 383–401.

Sidel, John, *Capital, Coercion, and Crime: Bossism in the Philippines*. Stanford: Stanford University Press, 1999.

Sikor, Thomas, "Agrarian Differentiation in Post-socialist Societies: Evidence from Three Upland Villages in North-Western Vietnam", *Development and Change* 32, no. 4 (2001): 923–49.

———, "Conflicting Concepts: Contested Land Relations in North-western Vietnam", *Conservation and Society* 2, no. 1 (2004): 75–96.

Sikor, Thomas and Thi Tuong Vi Pham, "The Dynamics of Commoditization in a Vietnamese Uplands Village, 1980–2000", *Journal of Agrarian Change* 5, no. 3 (2005): 405–28.

Sikor, Thomas and Tran Ngoc Thanh, "Exclusive versus Inclusive Devolution in Forest Management: Insights from Forest Land Allocation in Vietnam's Central Highlands", *Land Use Policy* 24 (2007): 644–53.

Sitthirith, Mak, "Territoriality of the Tonle Sap Lake and Implications for Resource Management". Paper presented at Southeast Asian Geography Association Conference. Singapore: National Institute of Education, Nanyang Technological University, 2006.

Sjaastad, Espen and Ben Cousins, "Formalisation of Land Rights in the South: An Overview", *Land Use Policy* 26 (2008): 1–9.

Sombat, Chantornvong, "Local Godfathers in Thai Politics", in *Money and Power in Provincial Thailand*, ed. R. McVey. Singapore and Chiang Mai: ISEAS and Silkworm Books, 2000.

Stonich, Susan C. and Peter Vandergeest, "Violence, Environment, and Industrial Shrimp Farming", in *Violent Environments*, ed. Nancy Lee Peluso and Michael Watts. Ithaca: Cornell University Press, 2001.

Straits Times, "Palm Oil Prices Spells [*sic*] Misery", 23 Nov. 2008.

Sugunnasil, Wattana and Suthawan Sathirathai, "Coastal Communities, Mangrove Loss and Shrimp Farming: Social and Institutional Perspectives", in *Shrimp Farming and Mangrove Loss in Thailand*, ed. E.B. Barbier and S. Sathirathai. Cheltenham, UK: Edward Elgar, 2004.

Sunderlin, William D., Jeffrey Hatcher and Megan Liddle, "From Exclusion to Ownership? Challenges and Opportunities in Advancing Forest Tenure Reform". Washington, DC: Rights and Resources Initiative, 2008.

Sutton, Keith, "Agribusiness on a Grand Scale: FELDA's Sahabat Complex in East Malaysia", *Singapore Journal of Tropical Geography* 22, no. 1 (2001): 90–105.

Szuster, Brian, "Coastal Shrimp Farming in Thailand: Searching for Sustainability", in *Environment and Livelihoods in Tropical Coastal Zones*, ed. C.T. Hoanh, T.P. Tuong, J.W. Gowing and B. Hardy. Wallingford, UK: CAB International, 2006.

Szuster, Brian, François Molle, Mark Flaherty and Thippawal Srijantr, "Socio-economic and Environmental Implications of Inland Shrimp Farming in the Chao Phraya Delta", in *Thailand's Rice Bowl: Perspectives on Agricultural and Social Change in the Chao Phraya Delta*, ed. F. Molle and T. Srijantr. Bangkok: White Lotus, 2003.

Tan, Stan B.-H., "Coffee Frontiers in the Central Highlands of Vietnam: Networks of Connectivity", *Asia Pacific Viewpoint* 41, no. 1 (2000): 51–67.

Taylor, Philip, "Redressing Disadvantage or Re-arranging Inequality? Development Interventions and Local Responses in the Mekong Delta", in *Social Inequality in Vietnam and the Challenges to Reform*, ed. Philip Taylor. Singapore: Institute of Southeast Asian Studies, 2004.

Technical Working Group on Forestry & Environment, "Forest Cover Changes in Cambodia 2002–2006". Paper prepared for the Cambodia Development Cooperation Forum, 19–20 June 2007.

Thangphet, Sophon, "The Impact of Urbanisation on Local Resource Management: A Case Study of an Indigenous Irrigation Community in Northern Thailand". Urban Planning, University of Sydney, 1993.

Thanh Nien News, "Drought Adds to Vietnam's Power Woes", 21 June 2010.

Thanh Tran Ngoc and Thomas Sikor, "From Legal Acts to Actual Powers: Devolution and Property Rights in the Central Highlands of Vietnam", *Forest Policy and Economics* 8 (2006): 397–408.

To Xuan Phuc, "Producing a Land Market in the Uplands of Vietnam". Unpublished manuscript, Anthropology Department, University of Toronto, 2007.

Tokrisna, Ruangrai, "Analysis of Shrimp Farms' Use of Land", in *Shrimp Farming and Mangrove Loss in Thailand*, ed. E.B. Barbier and S. Sathirathai. Cheltenham, UK: Edward Elgar, 2004.

Tran Thi Thu Trang, "Social Differentiation Revisited: A Study of Rural Changes and Peasant Strategies in Vietnam", *Asia Pacific Viewpoint* 51, no. 1 (2010): 17–35.

Trung Dang Dinh, "Coffee Production, Social Stratification and Poverty in a Vietnamese Central Highland Community", in *Understanding Poverty in Vietnam and the Philippines: Concepts and Context*, ed. R. De Koninck, J. Lamarre and B. Gendron. Laval, QC: Localized Poverty Reduction in Vietnam Project, 2003.

Tsing, Anna Lowenhaupt, "Agrarian Allegory and Global Futures", in *Nature in the Global South: Environmental Projects in South and Southeast Asia*, ed. P. Greenough and A.L. Tsing. Durham: Duke University Press, 2003.

Tubtim, Nattaya and Philip Hirsch, "Common Property as Enclosure: A Case Study of a Backswamp in Southern Laos", *Society and Natural Resources* 18, no. 1 (2005): 41–60.

Tulod-Peteros, Caroliza, *Golf Courses: Are They on a Par with Human Rights? Development and Human Rights: Series 1*. New Manila, Quezon City: Philippine Human Rights Information Center (PhilRights), 1999.

Tyler, Stephen, ed., *Communities, Livelihoods and Natural Resources: Action Research and Policy Change in Asia*. Ottawa: ITDG Publishing, 2006.

Vahari, Mehrdad, "An Introduction to Destructive Coordination", *American Journal of Economics and Sociology* 68, no. 2 (2009): 353–86.

Vandergeest, Peter, "Land to Some Tillers: Development-Induced Displacement in Laos", *International Social Science Journal* 55, no. 1 (2003a): 47–56.

————, "Racialization and Citizenship in Thai Forest Politics", *Society and Natural Resources* 16 (2003b): 19–37.

Vasudevan, Alex, Colin McFarlane and Alex Jeffrey, "Spaces of Enclosure", *Geoforum* 39 (2008): 1641–6.

Verdery, Katherine, *The Vanishing Hectare: Property and Value in Postsocialist Transylvania*. Ithaca: Cornell University Press, 2003.

Vidal, John, "The Great Green Land Grab", *Guardian*, 13 Feb. 2008.

Wakker, Eric, "Greasy Palms: The Social and Ecological Impacts of Large-scale Oil Palm Plantation Development in Southeast Asia". London: Friends of the Earth, 2005.

————, "The Kalimantan Border Oil Palm Mega-Project". Friends of the Earth Netherlands and Swedish Society for Nature Conservation (SSNC), 2006.

Walker, Andrew, "The 'Karen Consensus': Ethnic Politics and Resource Use Legitimacy in Northern Thailand", *Asian Ethnicity* 2, no. 2 (2001): 145–62.

————, "Seeing Farmers for the Trees: Community Forestry and the Arborealisation of Agriculture in Northern Thailand", *Asia Pacific Viewpoint* 45, no. 3 (2004): 311–24.

————, "Beyond the Rural Betrayal: Lessons from the Thaksin Era for the Mekong Region". Paper presented at the International Conference on Critical Transitions in the Mekong Region. Chiang Mai, Thailand, 29–31 Jan. 2007.

__________, "'Now the Companies Have Come': Local Values and Contract Farming in Northern Thailand", in *Agrarian Angst and Rural Resistance in Contemporary Southeast Asia*, ed. D. Caoutte and S. Turner. London and New York: Routledge, 2009.

Walker, Andrew and Nicholas Farrelly, "Northern Thailand's Specter of Eviction", *Critical Asian Studies* 40, no. 3 (2008): 373–97.

Warren, Carol, "Mapping Common Futures: Customary Communities, NGOs and the State in Indonesia's Reform Era", *Development and Change* 36, no. 1 (2005): 49–73.

Webber, Michael, "The Places of Primitive Accumulation in Rural China", *Economic Geography* 84, no. 4 (2008): 395–421.

Webster, Douglas, *On the Edge: Shaping the Future of Peri-urban East Asia*. Stanford: Asia-Pacific Research Center, Stanford University, 2002.

Wenk, Irina, "Land Titling in Perspective: Indigenous-Settler Relations and Territorialization on a Southern Philippine Frontier", in *Colonization and Conflict: A Comparative Study of Contemporary Settlement Frontiers in South and Southeast Asia*, ed. D. Geiger. Münster: LIT Verlag, forthcoming.

White, Benjamin, "'Agricultural Involution' and Its Critics: Twenty Years After", *Bulletin of Concerned Asian Scholars* 15, no. 2 (1983): 18–31.

__________, "Problems in the Empirical Analysis of Agrarian Differentiation", in *Agrarian Transformations: Local Processes and the State in Southeast Asia*, ed. Gillian Hart, Andrew Turton and Ben White. Berkeley and Los Angeles: University of California Press, 1989.

__________, "Economic Diversification and Agrarian Change in Rural Java, 1900–1990", in *In the Shadow of Agriculture*, ed. B. White, P. Alexander and P. Boomgaard. Amsterdam: Royal Tropical Institute, 1991.

__________, "Rice Harvesting and Social Change in Java: An Unfinished Debate", *The Asian Pacific Journal of Anthropology* 1, no. 1 (2000): 79–102.

White, Benjamin and Gunawan Wiradi, "Agrarian and Non-Agrarian Bases of Inequality in Nine Javanese Villages", in *Agrarian Transformations: Local Processes and the State in Southeast Asia*, ed. Gillian Hart, Andrew Turton and Ben White. Berkeley and Los Angeles: University of California Press, 1989.

Williamson, Peter and Philip Hirsch, "Tourism Development and Social Differentiation in Koh Samui", in *Uneven Development in Thailand*, ed. M.J.G. Parnwell. Aldershot: Avebury, 1996.

Winichakul, Thongchai, *Siam Mapped: A History of the Geo-Body of a Nation*. Honolulu: University of Hawaii Press, 1994.

Winkels, Alexandra, "Rural In-migration and Global Trade: Managing the Risks of Coffee Farming in the Central Highlands of Vietnam", *Mountain Research and Development* 28, no. 1 (2008): 32–40.

Winter, Tim, "Rethinking Tourism in Asia", *Annals of Tourism Research* 34, no. 1 (2007): 27–44.

Wittayapak, Chusak, "History and Geography of Identifications Related to Resource Conflicts and Ethnic Violence in Northern Thailand", *Asia-Pacific Viewpoint* 49, no. 1 (2008): 111–27.

Wolters, Willem G., "The Development of Property Rights to Land in the Philippines, 1850–1930", in *Property Rights and Economic Development: Land and Natural Resources in Southeast Asia and Oceania*, ed. T. van Meijl and F. von Benda-Beckmann. London and New York: Kegan Paul International, 1999.

Wong, Tim, "Communities on the Edge: Conflicts over Community Tourism in Thailand", in *Tourism at the Grassroots: Villagers and Visitors in the Asia-Pacific*, ed. J. Connell and B. Rugendyke. London: Routledge, 2006.

Wood, Ellen Meiksins, *The Origin of Capitalism: A Longer View*. London and New York: Verso, 2002.

Woon Weng-Chuen and Haron Norini, "Trends in Malaysian Forest Policy", *Policy Trends Report* (2002): 12–28.

Wyatt, Andrew, "Infrastructure Development and BOOT in Laos and Vietnam: A Case Study of Collective Action and Risk in Transitional Developing Economies", School of Geosciences, University of Sydney, 2004.

Yap, Emanuel, "The Environment and Local Initiatives in Southern Negros", in *The Politics of Environment in Southeast Asia: Resources and Resistance*, ed. P. Hirsch and C. Warren. New York: Routledge, 1998.

Yee Keong Choy, "Sustainable Development: An Institutional Enclave (With Special Reference to the Bakun Dam-Induced Development Strategy in Malaysia)", *Journal of Economic Issues* 39, no. 4 (2005): 951–71.

Yen Cao Thi Thu, "Towards Sustainability of Vietnam's Large Dams: Resettlement in Hydropower Projects". Department of Infrastructure, Royal Institute of Technology, Stockholm, 2003.

Zerner, Charles, *People, Plants and Justice: The Politics of Nature Conservation*. New York: Columbia University Press, 2000.

Zhang, Heather Xiaoquan, P. Mick Kelly, Catherine Locke, Alexandra Winkels and W. Neil Adger, "Migration in a Transitional Economy: Beyond the Planned and Spontaneous Dichotomy in Vietnam", *Geoforum* 37, no. 6 (2006): 1066–81.

Index

Printed and bound by CPI Group (UK) Ltd, Croydon, CR0 4YY

07/07/2026

14916220-0002